Cambridge Energy Studies

World Coal

World Coal

Economics, policies and prospects

RICHARD L. GORDON

Professor of Mineral Economics,
The Pennsylvania State University

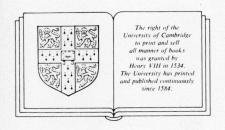

The right of the
University of Cambridge
to print and sell
all manner of books
was granted by
Henry VIII in 1534.
The University has printed
and published continuously
since 1584.

CAMBRIDGE UNIVERSITY PRESS

Cambridge
London   New York   New Rochelle
Melbourne   Sydney

Published by the Press Syndicate of the University of Cambridge
The Pitt Building, Trumpington Street, Cambridge CB2 1RP
32 East 57th Street, New York, NY 10022, USA
10 Stamford Road, Oakleigh, Melbourne 3166, Australia

First published 1987

Printed in Great Britain at the University Press, Cambridge

British Library cataloguing in publication data

Gordon, Richard L.
World coal: economics, policies and
prospects. – (Cambridge energy studies)
1. Coal trade
I. Title
333.8'22   HD9540.5

Library of Congress cataloguing in publication data

Gordon, Richard L., 1934–
World coal.
(Cambridge energy studies)
Bibliography.
Includes index.
1. Coal trade.   I. Title.   II. Series.
HD9540.5.G64   1987   338.2'724   86–26349

ISBN 0 521 30827 5

Contents

Tables

Figures

Foreword

This book is based on many years of work that has included examination as much as possible of the relevant literature and discussions with executives of coal and oil companies, electric utilities, government agencies, and international organizations. Here no effort is made either to report everything I have studied or to indicate everyone who has ever been of aid.

However, many new debts have been accumulated in the process of preparing for this book. First, Richard Eden of Cambridge University inspired this type of book with his invitation to contribute to the Cambridge Energy Studies series. He also provided much aid and hospitality along the way. A major element of the preparation for the book was arranging to speak with those concerned with coal in Western Europe, Australia, and South Africa. The South African trip was made possible by an invitation by Professor Theo Beukes to present a seminar at the Rand Africaans University. The Australian visit was made possible by a joint seminar with Professor Donald W. Barnett of Macquarie University. Professor Barnett visited me when the manuscript was nearing completion and made useful comments. Travel through Australia was assisted by Pennsylvania State's Australian–American Center. The visit to Europe was financed by pooling funds from the Texaco Foundation, Rocky Mountain Energy, Meridian Minerals, and the Matthew and Anne Wilson Fund of the College of Earth and Mineral Sciences of the Pennsylvania State University.

This book benefited from far more cooperation than any of my prior work. In particular, elaborate comments on the first draft were received from M. A. Adelman of the Massachusetts Institute of Technology, Vincent Calarco of Chase Manhattan Bank, Anthony Baker of IEA Coal Research, Ronald Steenblik of the London School of Economics, Peter Roberts of the Coventry Polytechnic Institute, and K. Reichert of the European Economic Community. Valuable, briefer comments were provided by Rudolph Schulten of Ruhrgas and Perry Phillips of British Petroleum. Calarco, Baker, Roberts and Reichert also made much useful material available. The suggestions were excellent and greatly contributed to the improvement of the book. What is now said is more correct and seems to me better expressed. The plethora of interesting suggestions about the need to add points to make discussions more satisfactory, in practice, led to scaling down of coverage. On many questions, I found no middle ground between a cursory mention of the basics and inappropriately long discussions of the extensive details of the quite different policies prevailing in each major producing country.

Many people provided aid during my trips abroad. Special appreciation for special assistance and hospitality go to Professor Eden, Walter Schulz, Dieter Schmitt, and Hans-Karl Schneider of the University of Cologne, Anthony Baker and his colleagues at IEA Coal Research, Peter Roberts, Peter Hughes and his associates at the Central Electricity Generating Board in the UK, John Jimeson and Robert Ovart at the OECD, Perry Phillips and colleagues at BP and his counterparts in Australia, Jerry Eyster and colleagues at Shell, London and his counterparts in Melbourne and Johannesburg, Ronald Steenblik at the London School of Economics, and Professor Max Gaskin. Others who were most helpful included Hugh Lee of the National Coal Board, Nigel Evans of Cambridge University, Cyril Lin and Michael Kaser of Oxford University, Jonathan Stern of the Royal Institute of International Affairs, K. J. Wigley and associates at the Department of Energy, Robert Deam of Queen Mary College, Dr. Bernhard Braubach and Dr. Vogel of the German Economics Ministry, S. Terjun of the Finance Ministry, Drs. Boke and Müller of the Bundesverband der Deutschen Industrie, Dr. Bernd Stecher of Gesamtverband des Deutschen Steinkohlenbergbaus, Friedrich Esser of Ruhrkohle, Werner Müller of VEBA, Gunther Meir of VEW, Herbert Troscher of RWE, Ernst-Otto Kantelberg of Verein Deutscher Kohlenimporteure, Erwin Zander of Hamburgische Electricitäts-Werke, Mr. Wachendorf of Vereinigung Deutscher Electrizitätswerke, Dr. Karlheinz Reichert and Kevin Leydon of the EEC, Pierre Simons of the Campine Mines, D. F. Anderson of the International Iron and Steel Institute, Walter Jensen of the Brussels Office of the National Coal Board, J. Verbesselt of the Association of European Coal Producers, Alain Bertay of Electricité de France, Marc Ippolito of Charbonnages de France, Jean Pelletier of ATIC, and Sven-Goran Rydahl of Cembureau. In South Africa, I was also able to visit with Anglo American, Gencor, Barlow Rand, JCI, and the Transvaal Coal Owners Association. In Australia, I met with both Broken Hill Proprietary and Utah International, CRA Australia, MIM, Coal and Allied Products, CSR, the national and New South Wales government energy ministries, and the Australian Coal Association.

Among those not visited who provided valuable material by mail were the UN Economic Commission for Europe, Saarbergwerke, the state of North-Rhine Westphalia, Badenwerk, Energie-versorgung Schwaben, the South of Scotland Electricity Board, ENEL, Österreichische Elektrizitäts-wirtschafts, the Irish Electricity Supply Board, the Swedish Syndkraft, and the Belgian economics ministry.

Formal permissions were sought from those sources not making clear what restrictions were imposed and generously granted for use of the data presented in the tables and figures. Those contacted included the OECD, Statistik der Kohlenwirtschaft, Verein Deutscher Kohlenimporteure, the (US) National Coal Association, the Australian Joint Coal Board, the Queensland Coal Board, the United Nations, the EEC, the South African Chamber of Mines, and Vincent Calarco. The tabulation and graphing process was facilitated by the appearance during the completion of this book of Microsoft's Excel program and the Macintosh Plus. All of the tables and graphs were generated with Excel. In many cases, the data handling capability of this type of program was used to lessen the possibility of error. Of course, these capabilities only ensure that once figures are correctly entered, further errors can be avoided when transferring the data to tables. Despite many checks, small errors in entering the original data keep appearing.

I am much better informed due to all this help and hope the book reflects the aid. As the discussion shows, the main regret is that too many governments are able to conceal critical data. The disservice to informed debate is a serious one.

My secretary, Theonas Fleming, once again endured my bewildering assortments of pasting together parts of several recent articles, micro-computer output, scribbles, and continued rearrangement and rewriting to make the material more coherent. Miraculously, she again made sense of it all. My wife, Nancy, found herself my companion on the South African and European trips and endured my absence in Australia and the time devoted to the book and many other activities.

Readers should note that thousands of documents were at least scanned to aid this research. These include statistical reports from many sources, policy studies from many international, national, and even regional groups, writings of outside observers, and a vast amount of material from coal and electric utility companies around the world. All this imposes formidable referencing problems.

I use the author date referencing system that relies on a bibliography to provide the basic references. This system seems the best to meet the needs of readers, authors, and publishers. The provision of as comprehensive a bibliography as possible appears vital in any book of this type. Once such a bibliography is available, the author date system reduces references to what should be clear direction to the germane entry in the bibliography. I find these more convenient to prepare and use than the traditional system of elaborate footnotes; this is particularly true now that such notes often are themselves separated from the text.

Several special problems occur here. First, a great deal of what has been examined in the nearly thirty years in which I have been studying these issues has no direct relevance or even use as background for most readers. Few will care to know that I skim over 100 annual reports and other documents from individual US electric utilities. Second, many facts are reported in several places. To deal with these problems, I have made the references selective and phrased the citations to ease identification. Thus, a blanket citation of a periodic report implies that every issue contains germane material. Conversely, when an agency prepares many reports, the citations in the text of any one report include either the title or a few key words that will ease identification of the relevant entry in the bibliography.

Finally, reports and sometimes the responsible agency change names. To shorten the bibliography, in most cases, only the latest version was listed. Also, where the titles contain an element identifying the year covered (e.g., Coal Distribution 1983), the references omit this.

In the months since the manuscript was submitted, I have continued to receive and analyze data on coal. Only a few new data were deemed important enough to incorporate for their own sake; a few others were added in the course of correcting errors. Two important points to note are (1) that not only are additional data available, but revisions have been made in the most recent data cited here, and (2) all data tabulations relating to the EEC refer to its ten 1985 members.

I prepared my own index and tried to list all cited authors, but not every casual mention of other names. Since almost every entry concerns coal, the phrases used invariably relate to some specific subtopic about coal.

Abbreviations

AGIP	Azienda Generale Italiana Petroli
ATIC	Association Technique de l'Importation Charbonnière
BACT	Best Available Control Technology
BHP	Broken Hill Proprietary
BP	British Petroleum
Btu	British thermal units
CDF	Charbonnages de France
CEGB	Central Electricity Generating Board
CMEA	Council for Mutual Economic Assistance
DOE	Department of Energy
ECSC	European Coal and Steel Community
ECU	European Currency Unit
EDF	Electricité de France
EEI	Edison Electric Institute
EIA	Energy Information Administration
ELCOM	Electricity Commission of New South Wales
ENEL	Ente Nazionale per L'Energia Elettrica
ENI	Ente Nazionale Idrocarburi
ESCOM	Electricity Supply Commission (South Africa)
FPC	Federal Power Commission
GE	General Electric Company
GENCOR	General Mining Union Corporation
IEA	International Energy Agency
IISI	International Iron and Steel Institute
ISCOR	South African Iron and Steel Industrial Corporation
JCI	Johannesburg Consolidated Investment
LDC	Less Developed Countries
MW	Megawatts
NAC	Natal Associated Collieries
NCA	National Coal Association

NCB	National Coal Board
NEPA	National Environmental Policy Act
NSW	New South Wales
OECD	Organisation for Economic Co-operation and Development
OPEC	Organisation of Petroleum Exporting Countries
OTA	(US Congress) Office of Technology Assessment
PSD	Prevention of Significant Deterioration
RWE	Rheinisch-Westfälisches Elektrizitätswerk
TCE	Tonnes of Coal Equivalent
TCOA	Transvaal Coal Owners Association
TDM	Theiss-Dampier-Mitsui
TOE	Tonnes of Oil Equivalent
UMW	United Mine Workers
USBM	US Bureau of Mines
VEW	Vereinigte Elektrizitätswerke Westfalen

CHAPTER 1

Introduction – a perspective on coal

For many years, coal industry performance has fallen short of the elevated expectations of coal enthusiasts. Contentions, nevertheless, persist that coal use must increase enough that coal will resume dominance of world energy markets. These views reflect inadequate understanding of the forces affecting coal. These influences *will* permit continuation of world coal industry growth. However, preeminence will not be restored. As in the past, substantial international differences in performance will prevail. Some regions will grow; others, decline. This study seeks to provide an introduction to coal that supports the case that the industry will expand, but not as explosively as enthusiasts contend.

From the depletion of British forests in the sixteenth and seventeenth century to the middle twentieth century, coal was the principal fuel for industrial development and an essential ingredient in the industrial revolution. Coal availability interacted with an outburst of human ingenuity, the opening of the western hemisphere, and many other forces to produce the era of industrialization and growth in which we still live.

After World War II, a drastic deterioration occurred in the market position of coal. Oil and gas captured most of the industrial, railroad, commercial, and residential fuel markets and surpassed coal as energy sources in much of the world. Coal use was increasingly confined to electric power generation and coking coal for ironmaking. The former market grew rapidly. The latter was stagnant over most of the post 1945 period and then collapsed in North America and Western Europe.

Rhetorical excesses habitually mar discussions of coal. Even when coal was most beleaguered, many argued that the limited availability of oil and gas allowed only a brief reliance upon them. Coal inevitably would resume its role as the chief fuel. The sharp rises in world oil prices in 1973–74 greatly increased enthusiasm about coal. Subsequent developments have diminished, but not eliminated, the hopes. They remain indestructible. Oil and gas will someday drop in importance. However, the evidence suggests that, at the earliest, this will not occur until several decades into the twenty-first century. By then, an alternative to coal also may have arisen.

Overly optimistic arguments about coal reflect misunderstanding of the problems of transforming coal in the ground into economic resources. The coal must be extracted, (in many cases) processed, transported to market, and then utilised. The extent to which coal is used depends on the total costs of utilization. The nonfuel costs of the post-mining steps are higher for coal than for oil and gas. Thus, the price of coal itself must be low enough to offset these other costs. For these and other reasons discussed below, the observed large physical endowment of coal does not inevitably guarantee the return to dominance. Only part of that endowment is competitive.

This book seeks to provide a compact, comprehensible examination of the critical economic issues relating to coal. Given a long record of extensive governmental intervention in coal markets, public policy questions receive considerable attention. The treatment may provide students of public policy with interesting examples. However, inclusion is primarily because intervention is so critical to coal.

Intervention reflects the tension between optimism about coal and the barriers to attaining the anticipated success. Governments engage in policies to promote the development of coal mines and technologies to process and utilise coal. Taxes or charges on anticipated windfall profits also are widespread and difficult to remove when they prove excessive.

Governments also adopt, when deemed necessary, measures to mitigate both the undesirable apects of coal production and use and any inability of their coal industries to remain competitive. Thus, universally, policies are imposed to protect the health and safety of coal miners, control the land disturbance from coal mining, and limit air pollution from coal burning. Western Europe and Japan have used numerous trade restrictions, subsidies, and limits on fuel choice to limit the decline of indigenous coal production. Coal in the communist countries endures all the problems the economic system imposes.

The discussion begins in Chapter 2 with examination of the economics of coal production, transportation, and use. The material provides the basis for delineating basic economic principles for analysing coal markets. The best studied, most complex markets are those in the advanced industrialised countries belonging to the Organisation for Economic Co-operation and Development (OECD). Therefore, data on coal use in these countries comes next, in Chapter 3. Developments in coal production and trade are presented in Chapter 4. Review of the organization of the coal industry is included. In the next four chapters, critical points about four key regions are examined. The areas are (1) the Communist countries, (2) the

United States of America, (3) Western Europe, and (4) the newly emerging coal exporters such as Australia and South Africa.

In designing each chapter, the intent was to highlight the characteristics that are most critical to understanding the subject. Review of the Communist countries in Chapter 5 presents background on coal production and energy use and emphasises the uncertainties surrounding future developments. The views of the USA and Western Europe consider only public policy. This is done both because regulation is so critical an influence in both areas and because the other key information is covered in earlier chapters.

When treating the countries that have developed coal exports since the 1960s, examination of production and trade trends, industry organization, and public policy is required (Chapter 8). In Chapter 9, the material is combined and extended to provide a review of possible future coal developments. Chapter 10 presents a summary and conclusions.

The discussion attempts to indicate the diversity of coal industries in the critical countries. Substantial intrinsic differences prevail among countries in the economics of producing and using coal. The United States and China have market conditions that produced large coal industries that are likely to continue growing. In the Soviet Union, coal-industry prospects are less clear. Countries such as Australia and South Africa have smaller industries likely to maintain high rates of growth. Other producers such as those in Western Europe are unable to compete with rival suppliers. The economics, as noted, are altered by several types of public policy.

To keep the discussion manageable, much was excluded. One omission relates to regions covered. The less developed countries (LDC) are largely neglected. Of them, only India is a significant coal producer. Such evidence as was examined suggests the Indian industry is problem plagued and able to compete only domestically. The problems apparently are of types similar to those in some countries discussed. Chapter 8 notes prospects for LDC participation in world coal trade. Another neglect is technology and its possible improvement. Discussions are available at many different degrees of complexity.[1] Only a sketch of the most basic points is given here.

Finally, all subjects necessarily are given summary treatment. This book provides a selective view, not an examination of all coal developments. Neither the consumption nor production discussions provide an exhaustive history. Few of the many tabulations made can be presented. The discussions of industry organization and public policies are deliberately limited. Those issues that are critical, manageable, and familiar to me are stressed.

The US, for example, is used as the main example of problems of pollution control from coal; Western Europe, as the illustration of protectionist policies. No effort has been made to compare policy making in coal to that in other realms.

Many aspects of coal seem incapable of satisfactory treatment at any level except extreme generality or extreme detail. These include labour relations, mine health and safety, mine reclamation, water pollution from mines, mine subsidence, and air pollution control. In every case, the situation differs from country and in each country, highly specific forces are at work. In all the policy realms mentioned, national programs consist of establishing many highly specific rules.

CHAPTER 2

An introduction to coal in the energy market

This chapter begins with discussion of the nature of coal. Then the forces making coal a less attractive energy source than oil and gas are examined. Attention then turns to the more complex question of how properly to appraise fuel supply prospects.

The nature of coal

A family of solid fuels exists. These range from low-heat, high-moisture content peat to the high-heat, high-carbon content anthracite. Peat usually is ignored in discussions of coal. The remaining grades of solid fuels are usually subdivided into two broad categories. First are lignites or brown coals, the lowest grades of solid fuels with significant commercial use. The second group of higher grade coals are usually termed hard coals.[1]

The basic differences in characteristics are the result of the complex physical and chemical reactions that transformed waste vegetation into solid fuels. The extent to which waste is displaced by useable fuel depends upon the time elapsed; other characteristics such as degree of contamination with sulfur or other minerals depend on local conditions. In addition, the characteristics of coal received by the consumer are affected by mining, processing, and possibly exposure during transportation.

Mining methods, for example, differ in the extent to which they separate coal from surrounding material. Producers make trade-offs between the cost and benefits of preparing coal by washing and screening to eliminate impurities. The result is a coal with more useable energy per unit of weight. This lowers transportation costs and eases combustion.

The experience with coal

From the seventeenth century in England through the start of World War I, the coal industry spread and grew. Growth was particularly intense in the nineteenth century. In the last half of the nineteenth century, the world oil industry began developments that matured into formidable competition for coal. Disruption by two World Wars, postwar territorial transfers that shifted the control of coal resources, and the Great Depression of the 1930s long obscured the underlying trends.

After World War II, it became increasingly apparent that the economic availability of oil and natural gas was far greater than previously realised. In particular, large oil reserves were developed in the Middle East. The result was shifts to the use of oil and gas at the expense of coal.

The rises of oil prices of the 1970s eroded the advantage of oil and gas albeit to a much lesser extent than enthusiasts about coal believe. A major reason for this is that oil prices have not continued to rise. Coal use has become more advantageous but only to the largest users.

The underlying forces – end uses and how to serve them

The central consideration in all appraisals of coal is the amount of coal that can be produced at costs significantly below the prevailing prices of oil and gas. No one would use coal unless the delivered price were below the delivered price of oil. Every coal user endures extra investment and (nonfuel) operating costs in receiving, storing, and burning coal.

Coal is a heterogeneous mixture of fuel and waste that must be torn from the earth and is difficult to process into a more satisfactory fuel. Oil and gas, in contrast, flow out of the ground and through the processing system with far greater ease than can coal. Moreover, oil and gas, particularly the former, can be more readily and thus more cheaply transformed into more valuable specialised fuels and chemicals.

Coal mining and, in fact, all stages of the systematic conversion of coal resources into useful products involve difficulties because coal is less tractable than oil and gas. These problems include numerous environmental impacts and other problems that inspire government regulation. The solidity of coal is one major disadvantage; the other is its lower energy content per unit weight compared to oil. A standard rule of thumb in energy data presentation is that anthracite and bituminous have about 70 per cent of the heat content per tonne[2] as does the crude oil.

A coal/uranium comparison is more complex. Uranium provides far more heat per unit volume than does coal. However, no simple widely accepted method exists for weighing the other characteristics of using the two fuels. The prevailing debate on nuclear power, in fact, centers on how to evaluate its characteristics.

Mining problems

Coal and uranium must be moved at every stage while oil and gas either flow naturally or can be induced to do so. All solid materials mining involves the basic steps of (1) getting access to the material, (2) undertaking whatever is necessary in such processes as drilling and blasting to make the material removable, (3) actually removing, and (4) moving it where it can be shipped. A critical distinction prevails between surface and underground methods. In the former, the resource is totally uncovered, and then the other stages are conducted.[3]

Underground mining involves building and maintaining some system of access to the coal. The simpler access methods involve directly tunnelling to the coal horizontally or on a slope. Then the coal can be transported continuously by a train or conveyor belt. Alternatively, a vertical shaft may be necessary with the hoists to move the coal and its miners.

Similarly, numerous combinations of techniques are available for the actual extraction and transfer to the removal system. Many factors affect productivity and cost. Work by the US consulting firm NUS identified six key factors – (1) seam thickness, (2) depth of cover, (3) roof conditions, (4) floor conditions, (5) gassiness of the mine, and (6) slope of the seam – as major cost influences on underground mines (1977a, p. 9). It is more advantageous to exploit a thick, level, strongly supported seam.

Mining is eased because expensive roof support arrangements can be avoided. Pillars of coal could be left behind to provide the support. This room and pillar arrangement long dominated US mining. Mines with less favorable situations were forced to construct supports. However, extensive European efforts to lower mining costs led to development of more mechanised, more productive methods of erecting supports and extracting the coal. Under the proper circumstances, use of these techniques in the more favorable mining conditions prevailing in the United States can produce costs below those of using traditional US methods. Research intended to revive European coal may have proved more useful to rival producing countries. European coal companies, in fact, seem more optimistic about selling technology than about producing coal.

Other critical distinctions relate to systems on extracting and transporting coal. A steady movement has occurred from primitive digging with pickaxes to increasingly sophisticated mechanical devices designed to cut large amounts of coal and dump it into the transport system. Carts have been the traditional mode of transport. Animal drive has been replaced by mechanical systems. In addition, conveyor belts are also used, and experiments have been made to move a coal water slurry in pipelines in the mine.

Surface mining involves simpler technologies than does underground mining. The critical offset is that surface mining by definition involves removal of considerably more cover than does underground mining. Where the cover can be removed by heavy earth moving equipment, the cost per cubic metre of cover removed is much lower than that of building and maintaining an underground tunnel system. In addition, removal and transfer to the shipping terminal generally are cheaper with surface mining methods. When very high volumes of cover or special difficulties in material handling arise, the underground option becomes preferable. The critical ratios change over time as the technology and economics alter. Surface mining with ratios of 25 or 30 metres of cover per metre of coal have become feasible in parts of the United States.

The critical influences are the amount of cover that must be removed and the ease of removal of cover and coal. A large seam with shallow cover involves the lowest costs. However, surface mines can operate profitably under a wide range of terrains and output levels. US mines include multimillion ton per year western operations that operate in thick seams with shallow cover. However, many smaller mines operate in the hilly terrain of Appalachia.

Different combinations of equipment are appropriate for different circumstances. Larger surface mines typically remove cover with large earth lifting equipment called draglines, remove the coal with smaller power shovels, and load it into large off-highway trucks. At the other extreme, some smaller deposits can be exploited using conventional construction equipment. Flexibility in coal output is provided by the ability of those owning such equipment to move between coal mining and road building. This has long been important in the US Appalachian region and has also become a factor in Britain, particularly since legislation lessened restrictions on entry into surface mining.

Technical progress has contributed to lessening coal mining cost rises throughout the world. The critical difference among countries relate to the extent to which mining conditions offset technical advance.

Significant variations exist intertemporally and interspatially in productivity. In particular, one of the most readily available indicators of the comparative economics of coal production in different countries is the disparity in output per man shift. In underground mining, data from each country's mining reports indicate that Australian and US levels are about 10–12 tonnes per man shift while those in France and Belgium are only two tonnes.

Other major problems of coal mining include often tumultuous labour relations, worker health and safety, disturbance of land surface mining, water pollution by all types of mining, and subsidence of land above underground mines. While labour relation problems are widespread, much remains to be determined about their nature and cause. Employment prospects within the industry have often been unstable. Many industries have gone through extended contraction. Even where the national industry is growing, regional difficulties can arise. From the late 1970s, coal output in some eastern US states has been depressed while western output grew. The opening of low cost surface mines in Australia has placed pressures on older underground mines. Since mines often are located in isolated regions, reemployment of displaced workers can be difficult. Resistance to relocation may aggravate problems of finding alternative work, probably to a lesser extent than advocates of preservation of mining contend.

Another obvious problem is that underground mining is dangerous and exposes workers to conditions that can be injurious to health. Workers seek to alleviate these dangers through both negotiations with management and by securing legislation to regulate mining practices. Whether the latter is effective as unions believe is questionable. Proof that collective bargaining cannot solve the problem is unavailable.

The characteristics of trade unions differ among countries and even among regions within a country and over time. A unified national union, for example, may dominate as was the case in Britain until the 1984–85 strike caused a split. Conversely, different unions for different crafts may operate as in Australia. A mélange may develop as has become the case in the US. The United Mine Workers (UMW) have lost ground to miners that are un-unionised or organised by other unions. Part of the problem has been inability of the UMW to organise in the western states in which most of the expansion occurred. In addition, many non-UMW mines opened in the East. Since World War II, the UMW passed through at least four phases. Initial militance was succeeded by a long period of cooperating with management to promote industry recovery. The 1970s were characterised by turmoil with contract renewal negotiations regularly leading to strikes. As demand for UMW output weakened, a more conciliatory attitude developed.

The European unions became more militant as European demand declined. The British mine workers, who had previously limited themselves to local walkouts, began to conduct national strikes. However, the collapse of the 1984–85 strike and the split engendered in the union may have altered the situation. The British apparently represent an extreme case. German unions are considered less radical, perhaps because they are under lesser pressure (see Chapter 7). As a result of bad publicity from strikes in New South Wales in Australia, efforts are being made to resolve disputes without disrupting production.

Subsidence is another coal industry problem in which potentially redress could be secured by suing the mine owners or arranging in advance for compensation. Legislation to force payment is imposed because of fears the companies will default or pay too little.

Land disturbance by surface mining and water pollution do meet the criterion that justifies intervention – the effects are so diffuse that negotiation by injured parties is infeasible (see Chapter 6 for a fuller discussion). However, in the disturbance realm, actual policies tend to regulate both the diffused and the specific effects. The problems of the owner of the land actually disturbed by the land could, in principle, be resolved by negotiation with the mine operator. Again regulations, nevertheless, often control these impacts. Complex, region-specific policies that cannot be covered here apply to surface mine reclamation.

Coal processing

If economically justified, coal can be upgraded by crushing, screening, and washing. This produces a more compact, cheaper to burn fuel. The decision to clean is thus determined by whether the resulting savings in transportation and utilization are sufficient. Cleaning plants conventionally are closely integrated with mining. Larger mines wishing to produce cleaned coal have their own cleaning plants. Smaller mines can be served by a regional plant, usually operated by a mining company.

Coking techniques can turn coal into a purer solid, a gaseous fuel, and other products including both wastes and valuable chemicals. Historically, the two most important variants were coking and gasification. Coking stresses production of a strong, low sulfur solid fuel, but the associated gas and chemical output usually is captured and utilised. In gasification more gas is sought and the solid may have little or no value.

Developments in energy markets have radically altered the relative roles of these approaches. Oil and natural gas competition have tended to eliminate both primarily gasification-oriented processes and coking to produce fuel for residential, commercial, and general industrial use. Increasingly, the market has become limited to production of coke useable in pig-iron making.

This contraction in use of coal to produce other fuels contrasts radically with the predictions that *more* transformations would occur. A major element in coal research has been efforts to develop technologies to create more cheaply better fuels from coal.

Coal is also always more expensive to transport than oil and, overland, gas is also cheaper to transport.

Coal utilization

A coal-burning boiler to produce a given output must be larger than one burning oil and gas to overcome the combustion problems associated with coal. Coal receipt and handling facilities also are more complex than those for oil or gas. Pollution control also is a greater problem with coal than with oil or gas. The large solid matter content of coal creates solid waste and air pollution problems much greater than with oil. Sulfur pollution differences are more complex. Different coals and crude oils differ in sulfur content. The sulfur in crude oil tends to concentrate in the heavier products burned as industrial fuels. Thus, which fuel would produce more sulfur dioxide air pollution in the absence of regulation can differ with the coal and oil alternatives available. However, sulfur is more cheaply removed from oil than from coal. Sulfur pollution from coal can be reduced by shifting to lower sulfur coals, trapping the sulfur oxides after combustion and before discharge to the atmosphere, or both. Sulfur oxide and solid matter air pollution cause various imperfectly understood dangers to human health, plant life, and exteriors of buildings.

Problems of coal use occur in all of the many (often incompletely reported) different uses of energy, but interuse difficulties differ considerably.[4] At one extreme, coal has prospects only in the railroad portion of the transportation sector. The widely used diesel electric technology could be replaced by electrification. Coal could be used in the powerplants. The prospects for increased electrification are limited, and examination of data on railroad fuel use suggests that greater electrification of railroads would produce only small changes in coal use. Prospects for coal are better outside of transportation. However, most coal use is likely to occur in electric utilities and a few other large scale energy users.

The coking case

A key special case is that of ironmaking. Coke – a solid fuel manufactured from blends of coal – has unique chemical and physical properties that make it a key input to the principal technique for ironmaking. Coke is made from processing mixtures of coals that will agglomerate into coke and are free from contaminants, especially sulfur.

In this market, steel output has long grown slowly and declined in the middle 1970s. The amount of pig iron produced per tonne of finished steel product has decreased. More steel is made from scrap in electric furnaces. New fabrication techniques reduce the amount of raw steel needed to produce a given volume of steel products. Unit use of coke per tonne of pig iron also has fallen. Use of other fuels and greater processing of ores before their injection into the blast furnace helped lower coke use. The result has been stagnation or worse in the only market in which coal has a technological advantage.

This is but one aspect of developments that have alleviated fears about coking-coal supply. Supplies are seen as unprofitably ample. Methods to avoid use of coking coal have long been known, and various newer alternatives have been proposed. These include making coke from lower grades of coal and adopting ironmaking processes not requiring coke. Some use is made of these alternatives. However, they are secondary to changing markets in lessening concerns about coke supply.

Another influence has been modification of coking practices to reduce use of the highest quality, most expensive coking coals. Fears about coking-coal depletion have given way to recognition that too many coal resources have been dedicated to future coking use. Thus, steel companies in the United States have greatly reduced their coal reserve holdings.

Given the special needs for the metallurgical coke market, coal output and sales are conventionally divided between coking and steam coals. Not all the coal that could be used in coking is actually employed in that fashion. Much finds its way into steam coal applications. However, compilers of production and export data that do not specify customers may report coal actually used as steam coal as metallurgical coal because it has coking properties.

Other uses

A direct or indirect method exists to serve all nonspecialised fuel needs through coal. Coal can be burned in boilers, and boilers come in many sizes to serve every combustion need. Conversion of coal into electricity and then utilizing the electricity also can serve most energy needs.

The economics greatly restrict the options actually adopted. A boiler is often so disadvantageous an alternative that it is rarely used. Thus, steam locomotion never succeeded in road transportation and became uneconomic in railroading. Even where a boiler is used, the drawbacks of a coal boiler compared to one using oil or gas often preclude coal use.

The extent of coal markets depends upon the interaction of user size and location. Location is critical for the usual reasons that it greatly affects transportation costs. Transportation, in turn, is a major influence on delivered prices. Thus, those with access to coal that is cheaper to mine and transport and further from sources of oil and gas are more likely to use coal. Coal use is most attractive in countries with ample low-cost coal supplies located near consuming centres.

The disadvantage of using coal can be materially reduced (but never eliminated) with large-scale burning. Major advantages are the ability to receive coal in trainload (or internationally in large shipload) lots and the lower unit costs of large-scale pollution-control equipment and its operation. Large-scale use produces lower unit non-fuel costs at almost every stage of the coal consumption process.

The scale effect appears to be increasing in importance and is of greater impact than location in affecting coal use. Thus, attention here is first directed at discussing the impacts of large-scale use. Bulk shipping has become an increasingly important influence. The sixties saw the provision of unit train rates in the US; the seventies, the construction of ports in South Africa and Australia to handle larger colliers.

Air pollution regulation, particularly since the 1977 USA move (see Chapter 6) to a best available control technology (BACT) approach, has also greatly increased the advantage of large-scale use. Not only are the unit costs of constructing pollution control devices lower for large-scale users, but important economies of scale arise in operation. The close controls and frequent cleanings required to keep stack-gas scrubbers operational are less burdensome for large operations.

The most clear-cut example of how coal use is more attractive at a large scale is electric utilities in the United States. US utilities have stopped ordering oil or gas-fired plants while continuing to plan coal-fired plants. For utilities, the fuel saving outweighs the extra nonfuel costs of coal use.

For an existing plant, the cost disadvantage of coal consumption is greater than for a new operation because the required special facilities are more expensive when added on instead of being integrated from the start.

Conversions to coal have occurred generally when little or no additional investment was needed. This is possible when the plant was originally designed to burn coal, still possesses the special facilities needed to use coal, and is not subject to environmental regulations that require addition of expensive control facilities. (As noted above, boiler-design considerations limit conversions to boilers built to burn coal. Efforts, however, have been made to effect coal use in boilers designed for oil use by diluting it with oil or water. The economic success of such methods has not yet been established.) Another approach has been to develop alternatives such as a fluidised-bed boiler that could ease coal burning. The commercial feasibility of these technologies also remains unclear.

The disadvantages of coal use are inherent. The customers are sophisticated about the nature of the alternatives. The view that poor marketing is a critical problem is a delusion, so the sales promotions programs pushed by the French and British coal industries have dubious prospects.

However, the drawbacks of coal also can be exaggerated. Some writers on energy use in Communist and underdeveloped countries impute a causal relationship between heavy reliance on coal and high energy use per unit of total output. The evidence suggests that this is at best an over-simplification. Some shifts to oil, e.g., adopting diesel-electric locomotives, will lower fuel use. However, in US electricity generation, the best coal plants historically had lower heat inputs than the best oil units.[5] The technology existed and was used to lower fuel use and the units could maintain steadier rates of operation and avoid the losses involved in irregular operations. Such plant design and operation differences generally are more critical to thermal efficiency than the fuel choice.

A possible offset to the advantage of coal over oil and gas in new electric power plants is that nuclear power could be even cheaper. The state of nuclear power differs radically among countries. Some, notably France and Japan, have moved to heavy reliance on nuclear. Others, such as Britain and Germany, have tried to expand nuclear power while maintaining a large role for coal. Some countries, such as Australia, have explicitly rejected the nuclear option. In the United States, electric utility optimism about nuclear power in the 1965–75 period inspired nuclear projects that had major impacts into the 1980s. However, rising regulatory difficulties eliminated willingness in the US to start new nuclear ventures.

Electric-utility fuel use patterns thus depend primarily on the fuel economies with the additional impact from attitudes to nuclear power.[6] In the United States, regional disparities in growth and the availability of fuels have had a major, complex influence. Industrial, and more critically electric utility growth, has been far faster in regions, particularly the West South Central States, in which coal historically was more expensive to use in generation than natural gas. However, the economics have altered so that coal has become the preferable fuel for new electricity-generating capacity in all regions.

Many observers regularly argue that rises in world-oil prices would cause the next largest type of fuel user – large manufacturing plants – also to find coal preferable to oil and gas in new facilities around the world.[7]

However, little, if any, progress towards attaining this goal has occurred (see Chapter 3). The main exception has been the cement industry; its technology allows acceptance of the impurities in coal. Thus, the industry has shifted heavily to coal. The problem in other industries seems to be that the forecasters underestimated the nonfuel costs (particularly for pollution control) of burning coal at the manufacturing-plant scale and believed that oil prices would continually rise substantially. The 1980 level was taken as a base for further increases instead of at best a plateau at which prices would remain for many years or, as the events of the 1980s suggest, an excessive level from which retreat was likely (see Chapter 9 for further discussion).

Depressed output in the industries most likely to use coal may have aggravated the difficulties.

However, in a large part of the world, coal from low-cost suppliers is a cheaper fuel than oil and gas for new electric power plants. Prevailing economic forces have led to an increasing tendency of electric-power use to dominate coal markets. As noted, a critical further question is the comparative economics of coal and nuclear power.

Coal transportation techniques and problems

Proximity to coal is an important but not conclusive determinant of coal choice. Most coal transportation techniques are those used for any bulk solids. However, in electric power markets, trade-offs are possible from moving coal or the electricity generated from it. The coal powered station could be located anywhere from the pithead to near the final market. The choice of location depends upon the comparative economics of generation at alternative sites of moving coal rather than electricity. A site distant from markets is likely to have the advantages of lower land costs and less severe environmental restrictions. Economies of scale may be attained by building one plant to serve several areas instead of one smaller plant for each area.

The comparative economics of transportation and distribution depend upon distance, volume, terrain, and the availability of facilities. Until congestion arises, waterway transportation is less costly than railroads. A coal seam located away from existing transportation networks, and in terrain in which powerlines are cheaper to build than a rail line, would be a candidate for mine mouth operations. Plants located next to mines are common around the world.

Coal can be mixed with water, and the resulting slurry can be pipelined. Studies show that adding and removing water imposes higher loading and unloading costs than railroads but these are offset by lower costs per tonne-kilometer of movement. Economies of scale also are important. The approach, therefore, is best for long distance, high volume hauls. The distance is necessary to produce enough of a saving on costs of movement to offset the higher loading and unloading costs. The volume is necessary to attain economies of scale.

Only one commercial coal slurry pipeline from a mine in Arizona to a power plant in Nevada, shipping about four million tonnes per year, operates.

The importance of different transportation methods is readily discernible only for the United States. The 1984 data are summarised in Table 2.1. Rail shipment is the most important method. The remainder is scattered among

Table 2.1 *US coal transportation by carrier in 1984*

Method	Thousand short tons	Percent USA shipments	Percent total
Rail to US customers	469,702	58.1	52.8
River	116,158	14.4	13.1
Great Lakes	13,532	1.7	1.5
Coastal ports	9,496	1.2	1.1
Trucks	111,263	13.8	12.5
Conveyors and pipeline	85,184	10.5	9.6
Other	3,051	0.4	0.3
Total domestic	808,386	100.0	90.9
		Percent exports	
Rail to Canada	1,074	1.3	0.1
Great Lakes shipments to Canada	17,779	22.0	2.0
Coastal ports	339	0.4	0.0
Truck	127	0.2	0.0
Total Canada	19,319	23.9	2.2
Overseas	61,410	76.1	6.9
Total export	80, 729	100.0	9.1
Discrepancy	349	NA	0.0
Coastal Ports	70,602	NA	7.9
Rail	59,838	NA	6.7
Truck	1,071	NA	0.1
River	902	NA	0.1
Unknown	61	NA	0.0
Great Lakes	31,311	NA	3.5
Rail	28,713	NA	3.2
Truck	48	NA	0.0
River	262	NA	0.0
Unknown	288	NA	0.0
		Percent USA	
Rivers	116,158	14.4	13.1
Rail	57,263	7.1	6.4
Truck	36,410	4.5	4.1
Conveyor belt	20,871	2.6	2.3
Unknown	1,914	0.2	0.2
Total shipments	889,462	NA	100.0

Source: US Department of Energy, *Coal Distribution 1984*

several other approaches often involving two transportation methods such as rail to ship or barge to ship.

The available evidence suggests that in western Europe a mix of rail and barge shipments and pit head power stations is involved. However, the waterway alternative is unavailable within such key areas as the Soviet Union, China, Australia, and South Africa.

Coal resources versus coal economics

Since coal seams often have portions that outcrop (i.e., are visible on the surface), occurrence is readily noted. Resulting efforts at further delineation produce estimates of substantial physical availability of coal. Discussions of coal often uncritically present such data in their introductory section. An immediate jump then is made to predictions of extensive increases in coal uses. The slogan that coal is the world's most abundant fuel is economically invalid and possibly questionable even as a physical concept. The ample amount of known coal resources is irrelevant. Mistaken comparisons of resource data for coal with reserve figures for oil has produced overoptimism about coal. The ultimate translation of minerals in the ground into reserves is an economic process. Neglect of this has led many into serious errors of overestimating coal prospects.

Economic abundance is cheapness in use. Such abundance often is quite different from large physical endowment. The forces discussed make it more attractive for most consumers to use oil and gas rather than coal but still leave enough customers for coal to permit output to grow. To see how these influences affect the economics of coal, it is necessary first to note the general economics of investment, turn to the application of this theory to mineral exploitation, and then show what this implies for fuel choice.

A note on the economics of investment

All the critical decisions in fuel supply and demand involve investments. Therefore, analysis requires understanding of the applicable economic theory of investment.[8] Interest charges must be imposed to insure an economically efficient allocation of resources. These charges, as do all other market prices, provide the best available indicators of relative attractiveness.

The disagreement is on whether actual market rates of interest are the most socially desirable rates. The debate has both an efficiency and a fairness component. Economic efficiency relates to whether the most advantageous use is made of resources given the prevailing ownership pattern. Fairness relates to whether that ownership pattern is desirable.

The efficiency debate is whether existing financial institutions can truly identify and finance the most attractive investments. Some fear that too few institutions exist to pool risks; others see remarkable abilities to meet rising needs. Observation of prevailing practices makes the latter view seem far more plausible. Conversely, critics of governments suggest they are biased towards quick results rather than to the farsightedness interventionists hope exists.

A correct statement of the argument, in fact, suggests that too much as well as too little attention can be given the future. Financial institutions produce costs as well as benefits. A financial process is only worth developing if the benefits exceed the costs. It is possible to suffer from the existence of devices that cost more than their contribution to better evaluation of the future. In fact, popular attacks on financiers shift as is expedient from charges of insufficient institutions to allegations that financiers promote excess dealings to fatten receipts. Another, highly intractable problem is the effect of both taxation and government spending (see e.g., Lund *et al.*, 1982). The critical fairness concern is whether market interest rates reflect the inadequate concern of present generations for future ones. All fairness arguments flounder on the lack of unambiguous bases for decision-making. What is right remains a highly subjective question. The one rule that is accepted is that generally transfers should go from richer to poorer. Economic growth makes future generations richer than present ones. Thus it may be unfair to adopt policies that treat the future better.

A general effort to treat the future better can be quite different from slowing the use of any natural resource (see Gordon 1966 and 1981). The best thing to do may be to more rapidly transform these resources into goods that allow higher living standards.

Arguments about the desirability of leaving more capital to future generations stress quantity over quality. The danger that much of the legacy will be obsolete is neglected. Thus, advocates of forcing interest rates below market determined levels fail to note such a policy has drawbacks as well as virtues. Moreover, lower rates may not lead to different results in the fuel choices that are the concern here.

However, the validity of arguments about the appropriateness of market rates of interest is secondary to the necessity for some rate of interest. The need to earn interest must be incorporated in all decisions. Future incomes must be discounted, i.e., reduced to account for the interest income lost by waiting.

Discounting at any reasonable rate causes only developments of the next few generations to affect current decisions. Thus, the relevant data for decision-making are those relating to resource availability prospects for those next few generations.

Coal and the economics of mineral resource development

In economics, shortages are considered a misnomer for supply problems. The proper concern is that market forces can cause sharp increases in price. More important than the terminology is the observation of practitioners of mineral resource economics that, at least in the absence of monopolies, mineral resource prices have been falling over time and these declines are likely to continue (see, e.g., Barnett and Morse, 1963, or Smith, 1979).

The argument here follows from a vision of mineral supply developed by M. A. Adelman (esp. 1970, 1972, and Adelman *et al.*, 1983). He recognises that, in the absence of additional investments, the costs of operating existing mineral properties will rise. Both the occurrence and the equipment will wear out. Deposit wear denotes such cost-increasing factors as the need to go deeper or farther from the center of operations or the declines in flow that occur as oil or gas production proceeds.

Producers move to another location in a deposit – horizontally to another bench or to a deeper bed in surface mines, or outward or downward from the original shaft in underground mining. These processes continue through the life of exploitation which in turn can proceed for periods ranging from the few months required to surface mine a small parcel of Appalachian coal to the many centuries for which some European metal mines have managed to persist.

A wide spectrum of further investment options is always available. In addition to refurbishing or extending existing operations, known but presently undeveloped properties can be activated, or new ones can be found and put into production. Developing and installing new technology is another option.

A classic error in mineral supply analysis is to overemphasise recent discoveries in augmenting supplies. In fact, enormous potential exists for increasing supply using the other options (including discoveries made long ago). These suffice to put considerable restraints on price increases. Conversely, new discoveries are only slowly converted into exploitable resources. It is less the immediate extraction of new discoveries than the development of past finds and the introduction of new technologies that holds down costs.

Exploration proceeds because of hopes that ultimately more profitable ways can be found to meet these demands. The continued success of such efforts explains both why exploration continues and why mineral prices do not rise sharply. Strictly defined, exploration only provides improved knowledge about mineral deposits. Too many writers ignore this and identify exploration with addition to reserves. In fact, such additions are not usually made until the discovery is developed into a producing property.

Moreover, exploration is a complex, difficult to define process. Resources are not simply known or unknown but range from those whose existence is unsuspected to those whose characteristics have been elaborately analysed. Generally, many favorable prospects are awaiting incentives before further delineation and development is attempted.

Establishing existence is not enough. Many factors such as ease of mining, processing, and transport and the size and quality of the occurrence affect its economic attractiveness. Information gathering must provide enough data on these characteristics to indicate where development will be profitable.

Adelman has argued that exploration produces what oil geologists call minerals in place – a stock of "known" but not necessarily currently useable minerals. As he stresses, the size of this stock is imperfectly known. Some of the minerals in place will not be economic to exploit immediately, and none can be exploited until producing facilities are created. Again following oil industry terminology, the stage of installing producing facilities can be termed development. Finally, comes the actual extraction and whatever further processing is required.

Exploration then at best provides tentative estimates of the magnitude of known minerals in place and, more critically, of their current economic viability. Even less is known about future economic attractiveness if only because this is so greatly affected by forces such as demand changes and technical progress that occur outside the deposit.

Moreover, the border between exploration and development is ill-defined. When activity takes place at the fringe of an established producing region, it is unclear and of no practical relevance whether this should be considered exploration or development.

Information always remains incomplete even after the minerals are produced and consumed. At no stage does it pay to discover every possibly relevant datum. The relevance of information, moreover, differs at different points. Investigation should be and is timed to coincide with needs. The magnitude of fact gathering is tailored to the extent of the investment involved. Less effort is made to justify leasing a property than to undertake basic reconnaissance.

As the firm moves on to even more costly activities such as extensive drilling and then development for production, increasing amounts of information are required. Even so, in every case, the effort will be incomplete. Information not immediately needed will not be acquired (at least unless a sufficient cost saving accrues to adding the effort on to the present program rather than undertaking a potentially much more expensive later effort).

Adelman (1972) showed that expenditure pattern prevailing in oil and gas stresses development and operation over exploration. While data are not as systematically compiled for other minerals, such indicators as discussions with those in the industry and regular reading of the trade press indicate his argument is probably even more relevant to solid minerals.

More generally, the mix maintained among development, exploration and other phases of mineral supply is a crucial indicator of expectations about expected costs. In particular, the more emphasis is placed on maintaining and extending operations and the less stress that is given finding new properties, the more optimistic we can be about supply prospects. Impending shortages will encourage extensive exploration. (However, an exploration boom can also be stimulated by realization of new opportunities to lower costs.) The more difficult it is to prevent cost rises by other means, the more likely that exploration will be stressed.

In fact, where pressures on price are so severe that it becomes profitable to hoard resources for future use, exploration will intensify to the point at which resources are being sought for stockpiling rather than for expeditious development. Absence of such efforts suggest exhaustion is not a pressing problem.

The problem in energy is not pending exhaustion, but rather the ability of certain oil-producing countries to exercise monopoly power to restrict output and raise prices. Given the limits to price increases imposed by alternative supplies and considerable opportunities to adopt energy-saving technologies, price rises due to monopoly cannot proceed unabated.

A major surprise of the high energy price era has been the extent to which it has proved effective to adopt energy saving technologies. This and rising non-OPEC supplies have put downward pressures on oil prices.

These falling oil prices suggest that those restricting oil supplies had adopted unsustainably large output reductions. At the very least, the minimum output goals of individual OPEC members summed to levels too high to be saleable at 1980 prices. It is less clear, but quite likely, that even if OPEC were a more cohesive cartel, it would have found it desirable to have allowed prices to fall from their 1980 levels to prevent severe losses from rising output by nonmembers and reduced energy consumption (see Chapter 9). Instability in the Middle East, moreover, may be more important because it allows price increases than because it disrupts supply (see Chapters 7 and 9).

Developing alternative sources of energy supply can and has contributed to weakening the OPEC country cartel. However, many supply options are available. The choice among them should be based on economic attractiveness rather than political appeal. Enough strong competitive suppliers of oil, gas, coal, uranium and nuclear reactors may exist that market forces will suffice to undercut OPEC. Any intervention to promote supply should be adopted with extreme caution.

Misunderstanding of these points produces fears of oil depletion. The slow growth in Middle East reserves is cited as evidence of impending supply problems. In fact, it is the natural response of countries attempting to restrict supplies. The potential to develop reserves remains great. Exploiting that potential would further undermine prices. Thus, the slowdown in reserve additions is the effect, not the cause, of high oil prices.

The superior use argument

The prior points about the economics of producing and using different fuels reinforce the argument about the attractiveness of continued oil and gas use. The economics imply that oil and gas constitute a high proportion of the most economic to use portion of at least the known stock of fuels. It is always preferable to use the most economic resources first. Waiting involves losing the opportunity to invest and earn interest on the cost savings.

Some commentators present the case for coal in the terms of a need to save oil and gas for superior uses such as (depending upon the speaker) petrochemicals or motor fuels. By definition, the absence of the emergence of such reservations of oil and gas implies that coal has failed the best-available test of superiority – that clear profits accrue to saving oil and gas for future use. (Comparable arguments exist about the undesirability of using cokable coals for steam raising.)

Here too the error is caused by neglect of interest rate considerations. The superiority of a particular use must be sufficient to repay the interest charges incurred in saving for meeting the future demands.

The concept of superior uses appears merely to restate the more familiar proposition criticised above, namely that inadequate effort is being made to plan for the impending rapid rises of oil and gas prices. As our experience with coking coal suggests, technical changes may make such savings unnecessary.

Resource data limitation

A more widespread, but less critical, mistake is misinterpretation of the available data on ultimately recoverable resources of different fuels. Radical differences prevail among estimates about coal, oil and gas, and uranium. These differences, moreover, also are explicable by the same arguments used to explain the supply picture.

Another consequence of the underlying economics is serious imperfections in the data on economic resources. Differences in the costs and benefits of gathering data on the availability of resources cause use of quite different techniques to produce available figures. As argued above, exhaustive data on the economic attractiveness of resources are expensive to secure and are produced only when essential. This occurs only when production is imminent. The ability to estimate physical occurrence is both highly imperfect and radically different from fuel to fuel. As Kaufman (in Adelman *et al.*, 1983) has shown, oil and gas resource estimation techniques are imperfect. The occurrence of coal in large continuous seams, some of which outcrops (i.e., is visible on the surface), makes it easier to find than oil and gas.

However, nothing can be clearly deduced from this about the comparative long-run economics of coal, oil, and gas. We are not sure of even the direction of the errors in different estimates. Thus, the net effect is unknown. In short, belief in an eventual return to coal arises from a misinterpretation of available data. That proved oil and gas reserves were limited and that little firm information existed about the potential discoveries were taken as evidence of impending shortage. Actually, confidence about the ability to meet demands without sharply rising prices discouraged heavy investment in precise knowledge about the availability of oil and gas.

Error was increased by the equally false impression that much more is known about the economic availability of coal and that these data suggest that coal is a better bet. Actually few if any statistics exist on the availability of *economically recoverable* coal. What we usually have are estimates of the *physical* occurrence of coal. As the prior sought to show, such occurrence is quite different from economic viability.

Moreover, imperfections exist in the data. At least in the US, initial estimates for explored areas tend to be overly optimistic. Some areas remain incompletely unexplored. Apparently, incomplete survey has become the more critical defect of US coal resource data. Thus, the estimates of the physical occurrence of coal may understate the total. This problem probably is even worse for countries such as China and India where far less exploration has occurred. However, the problem can arise in long exploited regions. The British National Coal Board followed up on indications that prior views of the extent of the South Warwickshire field were invalid. Subsequent drilling disclosed additional resources making it possible both to extend the life of existing mines and develop a new three million tonne per year mine (National Coal Board, 1985).

The economic recoverability of physically available coal is poorly known. The US government has attempted to delineate the economically-recoverable portion of this coal, but Martin B. Zimmerman (1981 and in Adelman, *et al.* 1983) has convincingly shown that the techniques used lead to severe overstatement of the actual economic availability of coal. The inclusion criteria are insufficiently stringent. An offset would be provided by omissions just noted of poorly explored areas, but the net impact is unclear.

Conclusion: implications for the consumption of coal

The available evidence suggests that the best policy for the next few generations is continued heavy reliance on oil and gas with coal remaining another important energy source. What we know about their availability suffices to justify this optimism. The general economic advantages of using oil and gas instead of coal reinforce the desirability of emphasis on oil and gas. The differences in the advantages of using oil and gas in different end uses, the so-called existence of superior uses, do not alter the appraisal.

At best, we can conclude that some day the depletion of oil and gas will reach the point at which it becomes more economic to increase greatly reliance on these coal resources than to continue heavy dependence on oil and gas. By that distant time, something even better may have emerged.

Errors in coal forecasting are largely the inevitable result of basic mistakes in general energy forecasting. The key mistakes were primarily betting on the exhaustible-resource rather than the imperfect-cartel model of oil prices and thus predicting a steady rise in oil prices, underestimating the ability to reduce energy use by substitution of less energy using equipment, and understating the prospects for natural gas. The special problems for coal analysis have involved great differences in prospects for nuclear power in different countries and the difficulties faced by the world steel industry. These difficulties include declining output and changes in the regional distribution of production (see Chapter 9).

Competition from coal industries in other countries and from nuclear power created additional problems for some coal industries, particularly those in Western Europe. These industries proved incapable of withstanding competition from these sources. Rising oil prices proved no relief. Thus, the protectionist programs instituted during the era of falling oil prices have persisted.

However, in some parts of the world, stronger coal industries were able to be effective competitors in the market for electric-utility fuel and take advantage of the substantial market expansion prevailing for electricity. The critical countries in this group are the United States, South Africa, and Australia. These countries and Canada also secured export markets for coking coal. The overoptimism about coal had the perverse effect that the successes of these countries are often denigrated because they fell short of expectations.

Thus, several distinct coal situations prevail internationally (and intranationally as well). The coal industries of different regions range from the stably growing to the dying. Those of Australia, South Africa, western Canada, and the western United States largely became established since World War II. These are the most strongly growing industries. Those of Western Europe, Japan, and eastern Canada have endured protracted contractions. Public policy excessively perpetuated the industries instead of efficiently easing the pains of a rapid contraction (see Chapter 7). Various intermediate cases arise such as the eastern United States in which a perilous reconstruction occurred in the 1960s and 1970s.

CHAPTER 3

Coal consumption trends in OECD countries

As with general economic activity and use of other energy forms, the larger more industrialised economies dominate coal consumption. Since 1971, the majority of world coal use has been in the Communist countries.[1] Since 1977, China has been the world's largest consumer of coal. The United States, long the world's leading consumer of coal, now is second. The Soviet Union is the third largest coal consumer in the world. These three countries account for 58 percent of 1984 world coal consumption (British Petroleum 1985, p. 27).

Two other major blocks of consumers collectively account for over a quarter of the world total. The rest of the Communist block absorbed 14.4 percent in 1984 (see Chapter 5 for partial data) – i.e., less than the Soviet Union. The nineteen European countries belonging to the Organisation for Economic Co-operation and Development (OECD) absorbed 11 percent. Five other countries – India, South Africa, Japan, Australia, and Canada – account for the great bulk of the remainder, about 12 percent. Thus, these countries and regions constitute over 95 percent of the market, and the scattered remainder can largely be ignored.

In this book, emphasis is on the OECD countries.[2] They encompass in addition to the European members, the USA, Canada, Australia, Japan, and New Zealand. Information on the communist block is harder to obtain, and what is available and viewed below suggests that the issues need less review than those in the OECD countries. This chapter begins with caveats about data and then sketches trends in coal use.

Problems of analysing energy consumption

Energy "balance sheet" data are widely calculated and employed to suggest the nature of differences in patterns over time and space. The procedure is convenient and conveys many insights. However, considerable difficulty is involved in putting the data together, and the data have numerous flaws that should be recognised.

The fundamental drawback is that all methodologies weight different fuels only by their heat content. As suggested in Chapter 2, heat content is only one economic attribute of a fuel. A value-based weighting would be preferable, since the price reflects the combined evaluation of all economically relevant characteristics. In practice, it is possibly only indirectly and qualitatively to recognise the importance of attributes other than heat content. Value data are too difficult to obtain and add into the analysis at reasonable cost. In addition, significant technical problems, discussed in an appendix to this chapter, arise in designing a data system.

Balance sheet practice causes a special problem for coal analysts. The distinction between coking and steam coal consumption maintained on coal tonnage data is neglected in balance sheets.

Trends in coal use

The clearest thing that can be observed about OECD countries is the tendency of coal use to be concentrated in electricity generation. In almost every other sector coal use proved far less convenient, and oil and gas has displaced coal. As noted, coke remains critical to ironmaking, but severe weaknesses long prevailed in that market in most countries.

The twenty-four OECD countries display considerable differences in many aspects of their energy and coal use.[3] Among these are differences in the extent to which the prior generalisations are valid. The United States dwarfs all the other OECD countries as an energy and coal consumer (Tables 3.1 and 3.2 and Figures 3.1 and 3.2). Another third of the total occurs in Europe; the remainder in the Pacific member countries – predominantly Japan and Canada.

The USA, Japan, Germany, Canada, the United Kingdom, France, Italy, and Australia are the critical countries. All but Australia consume significantly more energy than any other OECD country. Australia uses less energy but consumes more coal than several of the major users of total energy and has developed a growing role as a coal exporter. OECD averages are largely determined by these countries. Many members have patterns radically different from the average.

Electricity generation is the dominant market for solid fuels in the OECD countries as a group (see Table 3.2 and Figures 3.1 and 3.2). The relative role of electric power as a market is largest by far in the United States and the United Kingdom. In Australia and OECD Europe as a whole, nevertheless, a majority of solid fuel consumption is for electricity generation. A low proportion of Japanese coal consumption is for electricity. The

Table 3.1 *OECD energy consumption in 1983 (million tonnes of coal equivalent)*

a. Energy use by fuel

Region	Total energy	Solid fuel	Petroleum	Gas	Nuclear	Hydro	Solid fuel percent of country total
OECD total	5,075.3	1,242.4	2,231.4	955.4	275.6	369.4	35.0
North America	2,769.8	711.6	1,097.7	651.7	114.8	194.0	36.7
OECD Europe	1,685.4	395.0	784.0	251.3	120.8	133.2	33.5
EEC	1,294.8	311.7	600.4	238.8	93.6	47.6	34.4
Canada	301.9	46.3	100.9	58.7	15.5	84.9	21.9
United States	2,467.9	665.3	996.8	593.0	99.3	109.1	38.5
Japan	499.5	88.5	304.4	35.3	40.0	31.3	25.3
Australia	103.4	44.9	40.2	14.2	0.0	4.1	62.0
New Zealand	17.2	2.4	5.1	2.9	0.0	6.8	19.6
Pacific	620.1	135.8	349.7	52.4	40.0	42.2	31.3
Austria	36.8	7.3	14.6	5.6	0.0	9.8	28.4
Belgium	56.5	12.9	25.2	10.4	7.7	0.4	32.6
Denmark	23.7	8.1	15.0	0.0	0.0	0.0	49.0
Finland	35.1	10.1	13.6	0.8	5.7	4.3	41.1
France	267.0	39.2	127.9	32.6	46.0	22.8	21.0
Germany	362.0	120.5	157.2	56.0	21.0	6.0	47.5
Greece	23.4	7.0	15.3	0.1	0.0	0.7	42.8
Iceland	2.0	0.1	0.7	0.0	0.0	1.2	5.1
Ireland	12.1	3.2	6.0	2.6	0.0	0.4	38.0
Italy	188.2	19.5	118.0	32.6	1.8	15.0	14.8
Luxembourg	4.2	1.8	1.5	0.4	0.0	0.1	61.7
Netherlands	82.5	7.4	30.9	42.5	1.1	0.0	12.8
Norway	36.0	2.3	11.3	1.4	0.0	22.7	9.1
Portugal	18.0	1.8	13.4	0.0	0.0	2.6	14.3
Spain	102.9	25.7	61.6	3.1	3.4	9.2	35.6
Sweden	69.5	9.5	25.8	0.0	13.1	20.6	19.5
Switzerland	36.6	1.5	18.1	1.6	5.0	11.6	5.7
Turkey	53.7	25.1	24.7	0.0	0.0	3.6	66.9
United Kingdom	275.2	92.0	103.5	61.7	15.9	2.1	47.8

b. Fuel use by sector

Region	Electric	Final	Industry	Steel	Other industry	Trans-portation	Other uses	Misc.	Electric percent of country
OECD total	1,799.0	3,619.0	1,174.6	200.8	973.7	1,066.3	1,255.6	456.9	35.4
North America	968.5	1,984.6	555.8	65.8	490.0	673.7	690.9	234.8	35.0
OECD Europe	589.9	1,217.0	430.4	81.1	349.3	295.0	452.2	148.8	35.0
EEC	431.5	944.0	331.0	63.6	267.5	227.9	357.2	113.4	33.3
Canada	132.3	198.4	72.3	8.2	64.1	49.4	70.7	27.4	43.8
United States	836.2	1,786.2	483.5	57.6	425.9	624.4	620.1	207.4	33.9
Japan	191.8	338.6	159.0	48.2	110.8	68.1	94.6	62.0	38.4
Australia	40.2	68.2	25.9	5.1	20.8	25.9	14.7	9.9	38.9
New Zealand	8.6	10.6	3.6	0.7	2.9	3.6	3.2	1.4	50.2
Pacific	240.6	417.4	188.4	54.0	134.4	97.6	112.6	73.3	38.8
Austria	13.5	26.8	8.5	2.7	5.9	6.6	9.9	3.8	36.8
Belgium	16.3	43.0	17.6	5.9	11.7	8.4	15.6	5.4	28.9
Denmark	7.6	19.4	3.6	0.2	3.4	4.9	10.4	1.1	32.0
Finland	12.0	27.1	13.1	1.2	11.9	4.5	8.9	1.9	34.1
France	97.6	188.9	65.9	11.3	54.5	48.4	68.3	23.3	36.5
Germany	126.9	264.0	95.9	21.4	74.5	58.8	101.8	29.0	35.1

Table 3.1 – *cont.*

Region	Electric	Final	Industry	Steel	Other industry	Trans- portation	Other uses	Misc.	Electric percent of country
Greece	8.7	16.5	5.4	0.4	5.0	6.3	4.2	1.7	37.2
Iceland	1.2	1.2	0.5	0.1	0.3	0.6	0.2	0.1	61.0
Ireland	3.9	9.0	2.8	0.0	2.8	2.3	3.8	0.8	32.0
Italy	57.4	141.3	53.3	10.2	43.1	36.1	47.6	16.3	30.5
Luxembourg	0.3	4.0	2.3	1.8	0.5	0.8	0.8	0.0	7.8
Netherlands	18.1	66.8	25.1	2.8	22.3	12.6	27.1	7.2	22.0
Norway	22.8	23.1	10.0	1.9	8.1	4.5	7.3	4.4	63.4
Portugal	5.8	12.6	5.5	0.3	5.2	3.9	2.9	2.2	32.1
Spain	39.4	70.0	28.3	6.4	21.9	22.3	14.8	12.5	38.3
Sweden	36.1	43.8	17.0	2.3	14.7	8.7	17.0	4.6	51.9
Switzerland	16.9	25.3	5.6	0.0	5.6	6.9	11.9	1.7	46.3
Turkey	10.7	43.2	10.8	2.6	8.2	9.1	22.3	4.1	20.0
United Kingdom	94.6	191.1	59.2	9.6	49.6	49.2	77.6	28.5	34.4

c. Selected percents of OECD total

Region	Total energy	Solid fuel	Total fuel use in electricity
OECD total	100.0	100.0	100.0
North America	54.6	57.3	53.8
OECD Europe	33.2	31.8	32.8
EEC	25.5	25.1	24.0
Canada	5.9	3.7	7.4
United States	48.6	53.6	46.5
Japan	9.8	7.1	10.7
Australia	2.0	3.6	2.2
New Zealand	0.3	0.2	0.5
Pacific	12.2	10.9	13.4
Austria	0.7	0.6	0.8
Belgium	1.1	1.0	0.9
Denmark	0.5	0.7	0.4
Finland	0.7	0.8	0.7
France	5.3	3.2	5.4
Germany	7.1	9.7	7.1
Greece	0.5	0.6	0.5
Iceland	0.0	0.0	0.1
Ireland	0.2	0.3	0.2
Italy	3.7	1.6	3.2
Luxembourg	0.1	0.1	0.0
Netherlands	1.6	0.6	1.0
Norway	0.7	0.2	1.3
Portugal	0.4	0.1	0.3
Spain	2.0	2.1	2.2
Sweden	1.4	0.8	2.0
Switzerland	0.7	0.1	0.9
Turkey	1.1	2.0	0.6
United Kingdom	5.4	7.4	5.3

Source: OECD, *Energy Balances in OECD Countries 1983*, Paris, 1985
Tonnes of oil equivalent in the source converted to coal equivalent by multiplication by 10/7.

Notes to Table 3.1 – *cont.*

The other sector is agriculture, commerce, residential, and unspecified.
Miscellaneous is nonenergy,conversion losses, energy sector use, and statistical discrepancies.
OECD estimates total availability as production plus net imports less inventory increases.
Final consumption and its components include the heat value of electricity generated.
Electricity here is gross input, rather than the heat loss figure used by OECD.

Table 3.2 *OECD coal consumption in 1983 (million tonnes of coal equivalent)*

Region	Output	Total use	Electric	Final	Industry	Steel	Other industry	Trans-port	Other uses	Coking and gasworks	Other steam
OECD total	1,188.4	1,242.4	793.8	399.8	297.5	131.4	166.0	0.4	150.7	195.7	252.9
North America	749.2	711.6	505.5	193.6	147.3	39.4	108.0	0.0	58.8	39.3	166.8
OECD Europe	319.7	395.0	233.5	144.7	92.7	51.5	41.2	0.4	68.4	86.9	74.7
EEC	257.8	311.7	207.1	91.2	65.0	41.0	23.9	0.0	39.5	74.0	30.6
Canada	46.7	46.3	27.3	18.5	18.4	5.3	13.1	0.0	0.6	5.7	13.3
United States	702.5	665.3	478.2	175.1	129.0	34.1	94.9	0.0	58.2	33.6	153.5
Japan	15.0	88.5	23.6	46.6	45.9	36.4	9.6	0.0	18.9	64.1	0.8
Australia	101.8	44.9	30.8	13.0	10.4	4.1	6.3	0.0	3.7	5.3	8.7
New Zealand	2.6	2.4	0.4	1.9	1.1	0.1	1.0	0.0	0.8	0.0	1.9
Pacific	119.5	135.8	54.9	61.4	57.4	40.6	16.8	0.0	23.5	69.5	11.4
Austria	3.2	7.3	1.5	5.4	2.7	1.9	0.8	0.0	3.0	2.3	3.5
Belgium	6.3	12.9	5.2	6.3	5.2	4.4	0.8	0.0	2.5	6.8	0.9
Denmark	0.6	8.1	7.3	1.2	0.4	0.0	0.4	0.0	0.5	0.0	0.8
Finland	6.1	10.1	1.6	7.9	5.3	0.5	4.8	0.0	3.2	0.0	8.5
France	18.3	39.2	22.1	15.6	11.8	7.6	4.2	0.0	5.3	12.0	5.1
Germany	124.9	120.5	85.2	31.1	24.3	14.7	9.6	0.0	10.9	31.0	4.2
Greece	5.7	7.0	5.7	1.3	1.2	0.2	1.1	0.0	0.1	0.0	1.3
Iceland	0.0	0.1	0.0	0.1	0.1	0.1	0.0	0.0	0.0	0.0	0.1
Ireland	1.9	3.2	0.8	2.4	0.3	0.0	0.3	0.0	2.1	0.0	2.4
Italy	2.3	19.5	8.1	8.6	7.4	5.0	2.4	0.0	4.0	8.7	2.6
Luxembourg	0.0	1.8	0.2	1.6	1.6	1.5	0.1	0.0	0.1	0.0	1.6
Netherlands	0.0	7.4	4.8	2.2	2.1	1.8	0.3	0.0	0.5	2.8	-0.3
Norway	1.5	2.3	0.0	2.3	1.6	0.8	0.8	0.0	0.7	0.4	1.9
Portugal	1.3	1.8	0.3	1.4	0.9	0.2	0.8	0.0	0.6	0.3	1.2
Spain	18.7	25.7	17.6	7.1	6.6	3.9	2.7	0.0	1.4	4.7	3.3
Sweden	6.5	9.5	1.0	8.2	6.4	1.2	5.2	0.0	2.1	1.6	7.0
Switzerland	0.9	1.5	0.1	1.3	0.7	0.0	0.6	0.0	0.7	0.0	1.3
Turkey	23.6	25.1	4.1	19.9	3.4	1.9	1.5	0.3	17.3	3.6	17.4
United Kingdom	97.7	92.0	67.6	21.0	10.7	5.8	4.8	0.0	13.7	12.5	11.9

Source: OECD, *Energy Balances of OECD Countries*, tonnes of oil equivalent data multiplied by 10/7,
except coking from OECD,*Energy Statistics*, tonne for tonne.
Other Steam is total less electric and coking and gas.
Other uses include statistical differences, the energy sector, agriculture, commerce,government, and residential.

steel industry is by far the largest coal market in Japan (and if we add in the energy losses in coking, iron and steel probably accounts for the majority of coal use).

For various reasons, many other countries do not use most of their coal for electricity generation. Several countries such as New Zealand, Austria, Iceland, and Norway still get the majority of their electricity from water. Sweden, Finland, and Switzerland have both hydro and nuclear account-

Figure 3.1   Total hard coal consumption in OECD countries 1950–83

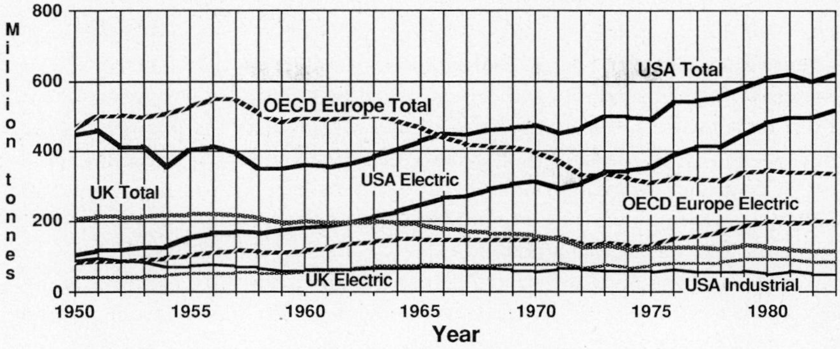

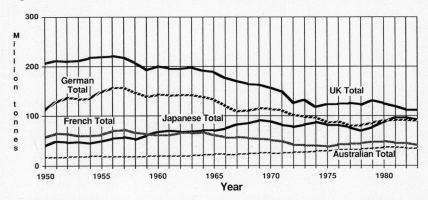

Figure 3.2   Total hard coal consumption in other OECD countries 1950–83

ing for a substantial portion of electricity. Turkey's coal goes mainly into miscellaneous uses, largely reflecting the dominance of such uses in all its energy consumption. Other countries – Italy, Spain, Turkey, Denmark, the Netherlands and Greece – have found it more attractive to shift from oil to coal. In Luxembourg, coal use by the steel industry is by far the largest market.

The exact coal market role of coking for pig iron manufacture is not readily deduced from energy balance sheet data. Therefore, the actual tonnes of input to coking are appended to the table.

Despite these problems, the basic picture does emerge. Japan is a far greater factor in the pig iron coking coal market than in the electric power and other markets; the reverse is true of the United States. Europe is a somewhat bigger factor in coking than in any other coal market. Within Europe, Italy, Belgium, and the Netherlands are notable for being bigger factors in the coking coal market than in other markets for coal.

Trends in electric utility coal use

The high relative role of electric power in coal use has been associated with a tendency towards rising coal consumption. The rise has not been universal and is by no means uninterrupted even in those countries with a trend to growth. The role of the US has been critical. Its position as the largest user by far of coal for electricity has involved steady growth in coal use. A tendency to growth has prevailed in Britain, Germany, Australia, Canada, and Spain. More irregularity prevailed in Japan, France, Italy, Belgium, and the Netherlands.

The steadiness of absolute growth of US electric utility coal use is not paralleled by stability of the coal share in generation. The share increased in the first two decades after World War II but started falling in 1966. The low of about 44 percent was hit in 1974; subsequent rises raised the share to 57 percent in 1985.

Growth of electric utility coal consumption was less persistent and universal outside the USA. Australia is the only country with steady growth. Canadian total solid fuel use in electricity did grow steadily. However, from 1971 to 1975, hard coal use dropped; the rapid growth in lignite use more than offset this. The bituminous drops were attributable to substantial increases in nuclear generation in Ontario.

Japanese coal use for electricity peaked in 1966, hit a low in 1974, and by 1983 remained well below 1967 (18 versus 26 million physical tonnes). Sharp increases in oil consumption prevailed from 1969 to 1975. Oil use was cut, but most of the displacement was by nuclear power.

For OECD Europe as a whole, a later more modest drop in coal use and a stronger recovery occurred. From 1971 to 1975, electric utility solid fuel use went from 185 million to 175 million tonnes of coal equivalent (TCE); by 1983 the level was up to 233 million TCE. Germany, the largest single user of solid fuel for electricity generation in OECD Europe, largely avoided the decline. Oil use rises were prevented by public policy (see Chapter 7). However, in the second largest market – Britain – coal use since the early 1970s has failed to grow. In France, electric utility hard coal use rose steadily from 1950 to 1964 from 11 to 18 million physical tonnes and stayed fairly stable until 1972 when several years of decline occurred. Growth then raised the burn to 22 million tonnes in 1983.

Italian coal consumption for electricity long has been modest and variable. Use never exceeded four million TCE from 1960 to 1978. Subsequently, the level has risen slightly reaching about eight million TCE by 1983. Spain surpasses Italy as an electric utility coal user. Spanish hard coal use fluctuated between one and two million physical tonnes of coal from 1950 to 1963 and then started climbing more regularly. Total solid fuel use in 1983 was about 18 million TCE. Denmark developed coal use in the 1970s as a conscious public policy.

Rapidly increasing supplies of oil and falling prices during the 1945–70 period encouraged increasing oil use in electricity generation. For various reasons including the completion of previously planned oil fired plants, oil use tended to rise well into the 1970s before plummeting. Some countries have had supplies of natural gas available. The US had an earlier and more substantial rise in gas generation than other countries. Markets for gas arose in some western European countries. Finally, nuclear power arose as a rival. The economic and political forces affecting these results differ radically among countries (see Chapter 9).

Coking coal declines

The behaviour of coking use of coal has differed even more radically among countries (see Figure 3.3). Many factors have arisen to hold down growth prospects. Nonsteel industry use has declined. As discussed in Chapter 2, steel industry technology has involved many developments that reduce the amount of coke needed per tonne of final output of steel products.

Coking coal use can rise only when steel output rises rapidly enough to offset declining coke use per tonne of steel. In the US, this was not the case. Up to about 1974, coking-coal consumption fluctuated greatly without growing. Depressed steel-industry conditions then caused a severe drop. The European situation is more complex. Use tended upwards in the first part of the 1950s to a 1957 peak of 161 million physical tonnes. A downward move set in, to about 127 million in 1974. Subsequently, declines have been sharper with only 87 million tonnes consumed in 1983.

Figure 3.3 Hard coal consumption for coking in OECD countries 1950–83

Within Europe, much variation occurred. In Italy and Spain, growth persisted into the middle 1970s. Italian consumption stayed stable since; a modest drop occurred in Spain. Of the three largest consumers, Germany and France have reduced levels somewhat less than Britain. At the 1957 peak, Germany used 73 million tonnes of coal for coking; Britain, 33 million and France, 25 million. The respective 1983 levels were 31, 13, and 12 million tonnes. Japan's growing steel industry increased its coking coal use through 1974.

Weaknesses in other markets

Beyond these difficulties is the failure of either other industrial markets or synthetic fuels markets to develop as optimists forecast (see Figure 3.4). The synfuels situation is less surprising. Basically, the case for synfuels was based on dubious projections about oil prices and overoptimism about synfuel costs. With continually rising oil prices, oil shale and various synthetics from coal could become competitive with oil.

The cause of the experience with other industrial markets is less clear. A standard forecast for over a decade has been for coal to absorb much of the growth in industrial fuel use. Numerous calculations were published purporting to prove that coal was the cheapest to use fuel source for new boilers. (The maturity of a newer method of combustion called fluidised bed reactor technology is creating optimism that the problem is close to solution. This remains to be seen.)

In the US, consumption through 1983 was stagnant in the other industry portion of the market. Sales never returned to 1973 levels. With US sales of

Figure 3.4   Industrial hard coal consumption in OECD countries 1950–83

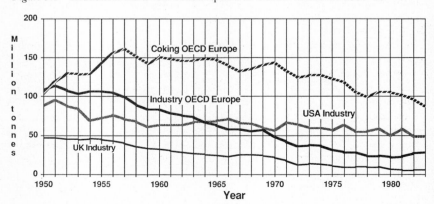

Industry includes energy sector

about 66 million tonnes in 1984, the first signs of a comeback appeared.

Similarly, in Western Europe and Japan a major shift to coal in noncoking industrial uses failed to emerge. Japan has raised coal use in this realm from only 1.5 million physical tonnes in 1973 to about 9 million tonnes in 1983. The long decline in European industrial coal consumption did not end until 1980 when the level was down to 19 million. By 1984, consumption of 27 million tonnes was still below the 1974 level of 32 million tonnes (and thus of the 86 million tonne level in 1955).

The expansion in electric utility coal use has sufficed to permit an increase in total US consumption of coal since 1960. In contrast, total western European coal use was declining through the late 1970s. By 1983, consumption was still well below post World War II peaks. Japanese consumption, in which coking coal dominates, had a more complex pattern – growth in the fifties and sixties, something of a downturn in the seventies, and a recovery in the early eighties. Thus, massive recapture of lost markets for coal failed to occur in the 1970s and early 1980s.

Appendix. The nature of coal and energy data

Numerous problems arise with coal and energy data. The two basic difficulties are the diffuseness of sources and the lack of uniform methodologies.

The diversity of publications is such that formidable difficulties arise in becoming aware of and locating all the germane material. Both international and national organisations provide data. The international scene is more straightforward. Each organisation usually reports on its members. The critical organisations here are the United Nations (UN), the Organisation for Economic Co-operation and Development (OECD), and the European Economic Community (EEC).[4] Each prepares various energy reports.

The UN presents a yearbook of energy data on all countries of the world. The main drawbacks are that only a few years are covered in each report and end use data are not provided.

The OECD maintains two principal energy data reports. One gives figures in natural units (such as tonnes of coal or oil and cubic metres of gas); the other presents tonnes of oil equivalent (TOE) data. At the minimum, reports fully covering the two most recent years on which data are available are issued annually. Reports covering longer spans are periodically issued. End uses are distinguished. The natural unit figures go back to 1950; the TOE data, to 1960. In 1985, balance sheets data on selected less developed countries were issued. The OECD has also undertaken many energy forecasts. In the middle 1970s, a special International Energy Agency (IEA) was set up in OECD. France, Iceland, and Finland declined to join so IEA is a smaller group than OECD. However, IEA is also the energy division of OECD.

Some IEA reports only cover its members, and others cover the full OECD. For example, annual reviews of overall energy policy and biennial reviews of coal policy only cover the IEA countries. An annual coal information report and irregular energy forecasts cover the whole OECD. The EEC prepares an energy yearbook (that has become less detailed over the years), a monthly energy bulletin (replacing separate ones for coal, hydrocarbons, and electricity), periodic energy forecasts, and an annual coal market review in its *Official Journal*.

A bewildering variety of national sources exists, and care must be taken to distinguish between original sources and rereporting. In the United States, the separate programs of the US Bureau of Mines (USBM) and the Federal Power Commission (FPC) have been consolidated in the Energy Information Administration (EIA) of the Department of Energy (DOE). It publishes numerous general energy, coal, and electric utility reports critical to coal analysts. Two trade associations, the National Coal Association (NCA) and the Edison Electric Institute (EEI), also present relevant data reports. NCA limits itself to compilation of federally generated data. EEI presents a mixture of US government data and material EEI collects directly. Prior to the creation of EIA, the EEI reports were far more satisfactory compilations of key FPC data than the Commission's own reports. However, EIA now issues a good summary of these data (and issues it before EEI publishes its report).

Elsewhere in the world, primary data are often generated by industry groups. Thus, the reports of nationalised coal and electric utility corporations in Britain and France are primary sources of much critical data. A private coal statistics bureau and the electric utility trade association in Germany compile critical data. The Belgian Coal Federation provides much useful data. The Joint Coal Board and the Queensland Coal Board provide critical data on Australia. The South African Chamber of Mines provides information on its members.

A useful part of the process is the effort by various groups to codify the material. The Ruhr coal owners association (Unternehmensverband Ruhrbergbau) produced at least three issues of a compendium of coal data for most critical countries. British Petroleum has long published a convenient annual compendium of data. Originally, only oil was covered, but the present format covers oil, gas, coal, nuclear, and hydroelectricity. The key data are neatly and quickly presented.

The speed of release and the extent of coverage differ widely among sources. At one extreme, a report may cover only one year; at the other it may attempt to provide the data for all years for which they exist. Unless the latter is done, compilation of data becomes difficult. In fact, it is almost certain that data from earlier reports will not be strictly comparable with the more recent ones. For example, the US EIA eventually revises all US energy consumption data back to 1960 (in the *State Energy Data Report*). However, a major redefinition of sectors discussed below means that there is no simple way to make pre-1960 data comparable with later ones.

Where data are not revised, compilers must make tradeoffs between convenience and accuracy. Ideally, one should start with the most recent reports for all they include and then consult the most recently issued report covering a prior year. Under such frequently used practices of including a specific number (e.g., 10, 20, or 30) of prior years, every earlier report should be examined. However, the temptation arises to undertake the less precise but more convenient route of viewing fewer reports and collecting several years data from each.

The sources differ considerably in both the way the end uses are defined and what measures are used. As noted, metallurgical coal is sometimes separated; sometimes it is aggregated. The IEA's work illustrates this. Coking is a sector in the natural unit report but not in the TOE data. At least five other treatments of coking coal have been used in the various IEA forecasts.

Another problem is in the techniques of creating balance sheets. Two main sources of ambiguity prevail in using heat content to develop energy balance sheets. First, different measuring rods are used; second, data on heat content are imperfect. The actual heat contents of fuels produced and consumed are not precisely measured. Three basic measuring rods exist. The British system concept is the British thermal unit or Btu (the heat needed to raise the temperature of one pound of water one degree fahrenheit). The calorie was long the preferred unit in energy data for metric system countries (the amount of heat needed to raise the temperature of one gram of water one degree centigrade). An alternative metric unit, the joule, has been adopted in many newer energy reports. (Its prime virtue is that with a simple shift of decimal points, Btu figures come close to those in joules). The rigorous definition (see any good dictionary) of a joule relates to concepts of work; however, it is also equal to a watt second or 1/3600 of a kilowatt-hour. The last is sometimes used as a measuring rod.

An alternative method of presentation is to display the data in terms of the number of units of a particular fuel – usually either coal or oil – that would supply the same amount of energy. Given the variability in heat contents of different coals and oils, the common procedure is to use rough, rounded figures. The most widespread convention is that a TCE contains 7,000 million (7×10^9) calories; oil, 10,000 million (10^{10}) calories.[5] Conversions from one measure to another is tedious (although much eased by the emergence of "spread sheet" programs for microcomputers) but creates presentation problems.

To reduce the conversion error, it is preferable not to round the numbers, but this creates an impression of spurious precision. A particularly difficult conversion is from barrels of oil per day. This measure often is provided showing fewer significant digits than other data. Here, I try to present either actual coal tonnes or tonnes of coal equivalent.

Imperfections also can prevail in both how the sectors are defined and how their fuel use is estimated. The creation in 1977 of a US Energy Information Administration (EIA) has produced extensive and still continuing efforts to reevaluate US procedures. These efforts have provided a particularly accessible demonstration of the limits of data quality.

At least one major conceptual change was made. Construction was moved from commercial to industrial to conform to the practice with other US data. Numerous prior estimation techniques were evaluated and found defective. Various shortcuts used by the agency previously in charge, the US Bureau of Mines, proved defective. For example, its coal consumption survey collected material on direct sales to manufacturing plants and on sales to resellers. The latter were presumed to sell only to residential and commercial users; resale to industry, in fact, also occurred.

CHAPTER 4

Trends in coal production and trade

As the prior chapters suggested, proximity to coal is a major, but not decisive, influence on coal consumption. Most hard coal and almost all lignite is consumed in the country of production. This chapter reviews the history of coal production and trade.

Basic coal industry characteristics are discussed, and production trends then are surveyed. Since few tabulations of solid fuel production in TOE or TCE terms are available, hard coal and lignite must be discussed separately. The history of international trade in coal is sketched. Finally, differential regional developments within several key countries are noted. Stress is on the most important producers (see Table 4.1).

Organising for coal production

Both how the coal firm is organised and how it relates to its customers has long been a concern of the industry, public policy makers, and outside observers. In evaluating industry structure, public policy makers may worry about whether the number of firms are too few to permit competition or too many to produce efficient operations. At different times in different countries, inadequate competition and excessive fragmentation were coal policy issues. Here, stress is on fragmentation.

The discussion throughout this book provides evidence that numerous firms in many countries are involved in coal production. These firms compete with each other and with producers of rival fuels. Capacity has expanded aggressively. To be sure, no conclusive practical method exists to appraise the vigor of competition, and critics of large corporations reject all efforts to conclude monopoly does not exist. Nevertheless, the evidence suggests vigorous competition prevails worldwide in the coal industry.

Whether the industry needs larger firms is more difficult to settle. A recurrent theme in the coal literature is that the industry is plagued with excessively small firms. Part of this argument is the usual implicit com-

Table 4.1 *Role of leading producers in coal world output 1984*

Country	Million tonnes of coal equivalent	Million hard coal tonnes	Million lignite tonnes	Percent of world total		
				Tonnes of coal equivalent	Hard coal tonnes	Lignite tonnes
USA	709.7	751.2	51.2	22.5	24.2	4.6
China	674.4	731.0	29.0	21.4	23.6	2.6
USSR	508.9	555.0	157.0	16.1	17.9	14.2
Poland	198.9	192.5	51.0	6.3	6.2	4.6
South Africa	126.0	162.0	0.0	4.0	5.2	0.0
India	140.0	144.0	8.5	4.4	4.6	0.8
West Germany	116.6	84.9	126.8	3.7	2.7	11.5
Australia	108.7	115.4	35.1	3.4	3.7	3.2
United Kingdom (1983)	100.1	116.4	0.0	3.2	3.8	0.0
East Germany	85.7	0.0	283.0	2.7	0.0	25.7
Czechoslovakia	71.3	26.4	102.9	2.3	0.9	9.3
Canada	51.7	46.2	9.4	1.6	1.5	0.9
Total Above	2,892.0	2,925.0	853.9	91.7	94.3	77.4
World	3,096.4	3,034.4	1,103.2			
World Adjusted for UK	3,153.7	3,101.5	1,103.2			

Source: British Petroleum, *BP Statistical Review of World Energy* (June 1985 edition), p. 26.
Converted to coal equivalent by multiplication by 10/7.
Except South Africa from Statistik der Kohlenwirtschaft,
Zahlen zur Kohlenwirstschaft, April 1985, p. 95.
BP reports total African output as 151.6 million tonnes of actual hard coal output
and 91.4 million tonnes of oil equivalent. The 10/7 factor implies only
130.6 million tonnes of coal equivalent. Average African output thus has about
78 percent the heat content of high quality coal. The 78 percent factor was
used to calculate the (high quality) coal equivalent of South African output .

plaint of industries and their apologists that competition is vigorous and monopoly profits cannot be secured. It is also contended that larger organisations would have the resources to undertake the planning, marketing, and research efforts needed to ensure "efficient" (i.e., greater) use of coal.

It was such arguments that led to the nationalisation of the British and French coal industries right after World War II and the consolidation of most of the Ruhr mines under a single company in 1968. As Chapter 7 shows, the specifics of the Ruhr case were quite different from those in Britain and France. The two last countries centralised to direct plans for industry expansion. Ruhr consolidation, offsetting prior policies of fragmenting the industry, was to facilitate contraction. The Ruhr merger, moreover, involved operations in one deposit and facilitated coordination among adjacent mines. Mines in Britain and France are more scattered than those in the Ruhr. US petroleum companies justified their moves into coal production as contributing badly needed improvements in management.

Experience suggests that the case for bigness was overstated. First, the coal technology problems discussed in Chapter 2 undoubtedly had far more to do with market difficulties than did industry structure. Second, corporations such as the British National Coal board and the German Ruhrkohle may have overcentralised the industry. Third, the infusion of management from outside has not clearly improved industry operation. Good (and bad) management can be developed in many ways. Fourth, differences in regional coal mining conditions and industry structure cause different structures to be needed in different places.

The United States has a particularly diverse and changing coal industry structure. The traditional pattern was for the industry to be dominated by independent firms of widely different sizes. The steel industry, notably US Steel and Bethlehem Steel, also were major producers as were a few electric utilities.

Many changes have occurred over the years. These include mergers between coal companies, purchases by oil companies and firms from various other industries, and the rise of new firms – again with a range of owners – independents, oil companies, electric utilities, and others.

The two giants of the US coal industry – Consolidation and Peabody – had their outputs peak in the 1960s. In both cases, the firms underwent management changes, albeit of very different types. Consolidation was acquired by Continental Oil, which in turn, was acquired much later by DuPont.

Peabody, which got its start supplying Samuel Insull's utilities and once was owned by Insull's Midwest Utilities, was acquired by Kennecott Copper. The Federal Trade Commission curiously considered this a more anticompetitive acquisition of a coal company than those by oil companies. The FTC successfully sued for divestiture, and Peabody passed into the control of a consortium including a large construction company, a life insurance company, and a firm with oil and gas interests. The lack of growth by both companies in the 1970s, however, seems more due to heavy involvement in large eastern underground mines than to their acquisition.

Sharply falling productivity due to new regulations and the influx of less experienced, more militant workers prevailed in US underground coal mining. Apparently the need to respond to these problems limited the ability to take the available opportunities for expansion. As noted below, the bulk of US coal output increase has been west of the Mississippi.

The growth companies have been those who participated in western expansion. This has been a diverse mix of companies. One old line coal producer, acquired by a major metals producer, Amax Coal, became a large factor. So did several electric utilities. One – Texas Utilities – recreated the Texas coal industry by opening lignite mines solely to serve its own plants. Utilities in other states developed enough capacity to become substantial sellers to other utilities.

A few other long-established eastern producers such as Westmoreland and North American developed western capacity. So did several oil companies such as Exxon, Mobil, Arco, Kerr McGee, and Sun. Construction companies such as Utah International, Morrison Knudsen (the last in partnership with Westmoreland), and Peter Kiewit also became established in the West.

In the East, many successes occurred – again by companies with diverse backgrounds – ranging from independents to oil company subsidiaries. In addition, disappointments, if not outright failures, occurred. Such companies as American Electric Power (the largest utility coal-user in the US) and Duke Power were severely criticised by regulators for having overly expensive coal operations. Charges of mismanagement were made. A more charitable explanation is that the utilities succumbed to the chronic error of anticipating substantial fuel price increases. The coal investments were made to protect the companies from these price increases. When prices failed to rise, the investments proved unprofitable.

The giant companies created in Britain and France did not produce the predicted renaissance, and Ruhr consolidation did not produce an expeditious contraction to efficient size. The key concern is whether the consolidations aggravated the problems. The British National Coal Board is often criticised as an oversized bureaucracy. However, it is not clear whether the key problem in Europe is company size or the politicisation of decision making.

The only clear-cut case is that, as discussed below, Soviet central planning is as harmful in coal as elsewhere.

Finally, the Australian and South African patterns are heavily influenced by local conditions (see Chapter 8 for more details). In both countries, large diversified companies dominate coal mining. In Australia, the leading mining companies are major factors. Broken Hill Proprietary is by far the largest producer. However, its holdings consist of mines it established and others acquired from US companies. Several other mining companies, a diversified company, a few companies whose main business is coal, Shell, British Petroleum, and the electric utilities of New South Wales and Victoria are other major actors.

The South African coal industry and much of the economy is dominated by what are termed mining houses. These firms started in metal mining but by mergers and expansion are enganged in a wide range of ventures. Their coal divisions were built around a core of existing, often previously independent companies. Shell, BP, Total (Compagnie Française des Petroles), and Ente Nazionale Idrocarburi (ENI) are also involved. In the first three cases, participation is through joint ventures with a mining house. State ownership occurs through the mines of the state owned iron and steel company. The synthetic fuel venture – Sasol – owns its mines.

Again local needs and historic accident seem more critical than an inherent need for big diversified companies as managers.

Optimal procurement arrangement

How best to arrange transactions is a specialised part of analytic economics, but a persistent concern of coal market participants. In 1937, Ronald Coase delineated the basic theory. He stressed that here, as in other economic choices, cost-benefit comparisons had to be made. Integration between producers and customers facilitated coordination and eliminated problems of regularly undertaking transactions. However, the extension of activities would stretch management and possibly raise costs. Contracts lessen the problems of making a deal. However, an ideal contract is impossible to develop. The uncertainty that necessitates contracts also prevents spelling out every possible development and what to do about it. Thus, contracts lessen, but do not eliminate, transactions problems.

Other theoretic writings have necessarily added little to Coase's conclusion that the right mix depends on the circumstances. Williamson (1975) excessively emphasises the disadvantages of open-market dealings and the desirability of integration to offset these difficulties. However, Stigler (1968) independently and tersely stated the other side of the case. He pointed out that when markets were large and well developed their use became less difficult and special arrangements were less essential. Coal experience shows that the variety of possible arrangements is extraordinarily diverse, that Coase's warnings are quite apt, and that Stigler is right: what is best depends on the extent of the market.

As the prior section noted, vertical integration does prevail between customers and coal producers. In Europe, coal producers have integrated into coke manufacture and electricity generation. Many other arrangements have been devised. US utilities have undertaken joint ventures, hired coal companies to operate mines, and bought reserves that they leased to a mining company. Contracts of many types have been signed. These range from ones agreeing to purchase output within broadly defined limits to agreements to repay investments no matter what occurs.

During the early 1970s, US coal producers and buyers thought that only "take or pay" contracts strongly committing the buyers to purchase or compensate would permit adequate capacity development. Unexpected cost rises in the early seventies forced considerable renegotiation. As costs stabilised and supplies proved more available, the necessity to contract seemed less. Moreover, it became increasingly apparent when a contract proves economically unsound, it is preferable to renegotiate it than insist on one's legal rights. The latter posture may only ruin the other participant. Similarly, reappraisals of contracts have arisen in Japanese purchases of coking coal. Contracts have allowed more flexibility and again adjust to changing circumstances. Contract clauses became less generous to producers.

The history of coal production

It is traditional, as noted, to date the start of a significant coal industry to the early seventeenth century. The nineteenth century was an era in which the previously much smaller coal industries of western Europe and the United States matured. Over the period, English predominance in coal diminished sharply as other countries established significant coal industries. As wood depletion began to influence the United States around the middle of the century, it turned to coal and by the end of the century became the world's largest producer – a position it has regularly maintained. Other major coal industries arose, notably in Germany.

While scattered data are available on nineteenth-century production, much more is available on the twentieth century.[1]

Twentieth century developments have been complex with many disruptions and reorganisations. The two world wars, the territorial changes produced afterwards, the great depression of the 1930s, the rise of competition from rival fuels, and the industrialisation of some countries have variously affected the trends.

At the start of the century, the United States and Britain together accounted for two thirds of world hard coal production (see Table 4.2 and Figures 4.1 and 4.3). Only a few countries (or more precisely only the present territories of a few countries) accounted for more than 2 percent of world output – Western Germany, (10.4 percent), France (4.8 percent), Poland (4.2 percent), Belgium (3.4 percent), and the Soviet Union (2.3 percent).

By the 1980s, the pattern had radically changed. The United States remained the leading producer but provided less than a quarter of the output. China's industry was producing only slightly less. The Soviet Union was the number three producer with less than a fifth of the world total. Poland had become number four producer. South Africa, India and Australia all had expanded greatly.

Western European countries had greatly reduced output. The coal

Table 4.2 *Twentieth-century coal production in selected countries and years (million tonnes)*

a. Selected leading hard coal producers

Year	World	USA	USSR	China	Austra-lia	Poland	South Africa	India	Canada
1900	701	245	16	NA	6	30	1	6	5
1913	1,216	517	30	16	13	49	8	16	13
1920	1,192	595	7	23	13	36	10	18	12
1929	1,325	550	37	25	11	74	13	24	12
1938	1,204	355	113	31	12	69	16	29	12
1946	1,215	537	139	11	14	47	23	30	12
1947	1,369	621	154	14	15	59	23	31	10
1948	1,405	593	170	9	15	70	24	31	12
1958	1,815	389	353	270	20	95	37	46	7
1960	1,983	392	375	420	22	104	38	53	7
1970	2,179	550	474	360	45	140	55	74	8
1978	2,600	574	557	593	72	193	90	102	17
1983	2,919	663	558	688	99	191	146	136	23
1984	3,018	743	555	740	114	192	162	145	32

b. Selected Western European hard coal producers

Year	Britain	Germany	France	Spain	Belgium	Nether-lands	Italy
1900	229	73	33	3	23	0	0
1913	292	132	44	4	23	2	0
1920	233	101	24	5	22	4	0
1929	262	145	54	7	27	12	0
1938	231	151	47	6	30	13	1
1946	193	62	47	11	23	8	1
1947	201	82	45	10	24	10	1
1948	213	100	43	10	27	11	1
1958	219	149	58	14	27	12	1
1960	197	142	56	14	22	12	1
1970	145	117	37	11	11	5	0
1978	122	90	20	12	7	0	0
1983	116	90	17	15	6	0	0
1984	49	85	17	15	6	0	0

c. Leading lignite producers

Year	World	West Germany	East Germany	Czecho-slovakia	USSR	USA
1938	262	68	120	16	19	3
1946	242	52	110	19	50	2
1947	268	59	102	22	51	3
1948	294	65	111	24	58	3
1958	619	94	215	57	143	2
1960	644	96	225	58	135	2
1970	796	108	261	82	145	5
1978	952	124	253	95	163	35
1983	1,088	124	278	102	158	47
1984	1,131	127	295	105	157	54

Sources: For all countries and all coal types:1900,1913,1920,1929,1946,1947,1948 and China 1938, 1958 and 1960: Unternehmensverband Ruhrbergbau, 1955 and 1961, *Die Kohlenwirtschaft der Welt in Zahlen.* 1937, 1958, and later years:Statistik der Kohlerwirtschaft, *Zahlen zur Kohlenwirtschaft* , various issues. Except Australian data from Joint Coal Board, *Black Coal in Australia* and USA lignite 1958 and 1960 from OECD, *Energy Statistics.* Pre World War II Poland has Silesia as reported by Germany.

Figure 4.1 Hard coal output of leading producers 1946–84

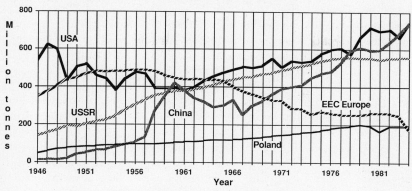

EEC Europe includes Spain but not Portugal

Figure 4.2 Hard coal production in other countries 1946–84

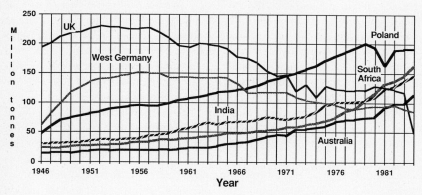

Figure 4.3 Output of leading lignite producers 1946–84

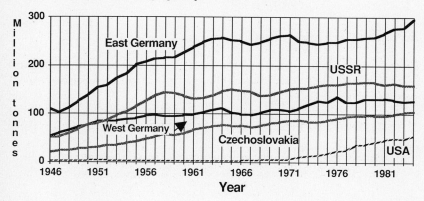

industry in the Netherlands had closed; French and Belgium had greatly decreased output. Lesser proportional but larger absolute declines arose in Britain and West Germany. As discussed further below, while the German industry is higher cost than the British, German decline was less than British (see Table 4.2 and Figure 4.2).

Many forces produced these trends – the two world wars, the great depression, the efforts to build up poorer countries, and the shift to other fuels. Countries differ radically in what impacts various developments had. Examining the data is necessary to determine the effects of the critical influences.

Lignite history is influenced by the same trends. The pattern of lignite output, however, differs considerably from that for hard coal. East Germany, the Soviet Union and West Germany have long been the leading lignite producers (see Figure 4.3). Although their share of world output has diminished, the three still produced in 1984 over half of world lignite output. Czechoslovakia was the only other country producing over 100 million tonnes of lignite. Generally lignite output and use occurs because of special local conditions. Except perhaps for those in the Soviet Union, most producing regions are ones either without hard coal resources or with difficulties in expanding hard coal output. This clearly is true of the two Germanies. Examination of the details for the United States and Australia shows that in both countries lignite output occurs in hard-coal poor regions – Texas and North Dakota in the US and Victoria in Australia.

In any case, addition of lignite merely adds a few countries to the list of middle rank producers. As Table 4.1 shows, the dominance of the USA, China, and USSR continues when lignite is added. West Germany proves a more important solid fuel producer; in particular, when lignite also is considered, it is a more important producer than Britain. Inclusion of lignite also increases the importance of East Germany and Czechoslovakia as solid fuel producers. However, they remain clearly small scale.

International coal trade patterns

Analysis of available data shows that coal use largely occurs in the country of production.[2] Only Australia has emerged as a major coal exporter whose exports exceed domestic consumption. However, many countries secure the bulk of their coal from imports. Japan is by far the most important example. The other main examples are in Western Europe (recall Table 3.2).

World coal trade has undergone significant changes over the years. Prior to World War I, the major components of international trade were intraEuropean flows principally involving British and German exports and US exports to Canada. France, the Netherlands, Italy, Germany, and Belgium were the principal import markets. The Netherlands and Belgium got the majority of their coal from Germany, and the other countries relied more on Britain. British and German exports declined sharply after World War I and then tended to decline further.

Both the recreation of an independent Poland after World War I and the territorial realignment after World War II provided the country with coal producing regions formerly in Germany. The markets, however, remained in Germany and elsewhere so Poland became a major exporter.

The United States gradually established an important position in overseas markets – first in western Europe and then in Japan. As the Soviet Union became a major producer, it too began to develop a substantial export trade – predominantly to the other Communist countries (see Chapter 5). Subsequently, Canada, Australia, and South Africa began to develop export markets. In the first two cases, the primary impetus was the ability to provide coking coal to the growing Japanese steel industry. In the Australian case, steam coal sales and European markets ultimately developed. South Africa's coal exports are predominantly of steam coal. Europe is the chief market, but some sales including a few million tonnes of coking coal go to Japan. (These coking coal sales had a critical historical role; they were the first significant exports.) (See Table 4.3 for summary data.)

These developments have created severe competitive pressures for the US which has been a major supplier of coking coals to western Europe and Japan and sells significant amounts of steam coal in western Europe. In fact, in 1984 Australia exported more coal than the United States, which had been the leading exporter for many years. The United States and Australia are the largest exporters and account for nearly half of exports. Poland and South Africa each account for about 10 percent of exports; Canada and the Soviet Union, 6 percent.

Western Europe is the largest importer of US coal; Japan and Canada are the other main markets. Both steam and coking coal moves to Europe and Canada; Japanese purchases mainly coking coals. Australia and Canada sell mostly to Japan. Canada has developed a significant market in Korea; Australia, an even bigger one in Europe. South Africa, as noted, sells mostly to Europe but with significant sales to Japan.

The Soviet Union still exports mainly within the block. Polish exports are more evenly divided between Eastern and Western Europe.

These patterns are quite different from those of 1913 or even 1959. Prior to World War I, the bulk of world trade took place inside western Europe. Britain and Germany were the main exporters. The importers were their neighbours. US exports to Canada were the other big element.

Table 4.3 *Selected data on world coal trade 1970, 1981 and 1984 (million tonnes)*

	1970	1981	1984
	a. Total		
USA	65.0	102.1	73.9
Australia	19.0	47.2	75.9
New South Wales	12.0	22.3	35.7
Queensland	7.0	24.9	40.1
Coking	16.9	37.4	47.0
Steam	2.1	9.8	28.8
South Africa per Chamber of Mines	1.3	29.9	38.1
South Africa per German Coal Importers	NA	29.5	36.7
Poland	28.8	15.0	43.1
Canada	4.0	16.3	24.4
	b. To Western Europe		
USA	19.8	51.2	28.4
Australia	2.2	7.3	16.8
South Africa	0.8	20.2	20.7
Poland	15.3	7.0	22.4
Canada	0.1	1.4	2.4
	c. To Japan		
USA	25.0	23.5	14.8
Australia	16.1	32.2	41.0
South Africa	0.3	4.6	7.8
Canada	3.7	10.9	15.9

Sources: USA, Canada, Poland, and South African data for 1981 and 1984 :
      Verein Deutscher Kohlenimporteure, *Jahresbericht 1984*.
      USA and Poland 1970 from 1975 edition. US may include some Eastern Europe.
      Australia: Section (a) data and 1984 regional details: Joint Coal Board, *Black Coal in Australia 1984-85*, pp.89-93.
      Regional details for fiscal 1981-82 from *Black Coal in Australia-1982-84*, p.83, used for 1981.
      Australia to Japan, fiscal 1970-71, for New South Wales from *Black Coal in Australia 1984-85*, p.103 and for Queensland from Queensland Coal Board,*34th Annual Report*, pp. 114-15, for 1970.
      Australia to Western Europe, fiscal 1970-71 from National Coal Association, *World Coal Trade 1972*,, p. 40, used for 1970. Canada 1970 and regional details for South Africa 1970 (calculated as sum of import data for counties reporting imports from South Africa): National Coal Association,*World Coal Trade 1972*.
      Alternative South African totals from Chamber of Mines, *Statistical Tables*.

Qualitatively, the patterns were similar in 1929; the volumes were lower. Poland had emerged as an exporter.

After World War II, the export ability of Britain and Germany diminished markedly. US and Polish exports rose. The US developed its European and Japanese markets. The former fluctuated sharply because public policy forced imports to bear the brunt of demand swings. It was not until the 1960s and 1970s that Australia, South Africa, and Canada emerged as exporters.

Changing intranational coal production patterns[3]

The patterns of coal production within different countries has altered radically over time. In the United States, Appalachia, the easternmost strip of coal resources, has always been the largest producing area. However, its role has diminished. Particularly pronounced changes prevailed from the 1970s on. Appalachian output grew in the sixties but stagnated through much of the seventies, rose a bit and declined again. The Illinois basin had even more severe problems. It expanded in the 1960s but has since declined.

In the meantime, various forces encouraged the rise of extensive coal production west of the Mississippi. Wyoming, Montana, and Texas enjoyed the greatest growth, but Colorado, Utah, Arizona, and New Mexico also expanded significantly.

The major impetus was the shift to coal for new plants by electric utilities west of the Mississippi. In addition, US air pollution regulations stimulated utilities in Minnesota, Wisconsin, Illinois, Indiana, Michigan, and Ohio to use low sulfur western coal.

In Europe, contraction of output has been concentrated on older, higher cost areas. Particularly striking changes occurred in Belgium and France. Belgium felt severe pressures to preserve coal employment in its very high cost southern fields. However, these have been closed, and the remaining output is in the Campine.[4]

The majority of French output through 1955 occurred in the Nord/Pas de Calais region. That region hit its postwar peak output of 29 million tonnes in 1952. Given subsequent contractions, 1984 output was down to 2.5 million tonnes. The seven small southern regions, collectively known as the Centre/Midi, followed a similar path. Output went from a peak of 15 million tonnes in 1958 to below 5 million tonnes in 1984. In contrast, Lorraine output went from 15 million tonnes in 1959 to over 10.9 million tonnes in 1984.

British output has become more concentrated in central England. In particular, mines in Derbyshire and Nottingham have increased output share from 15 percent in 1946 to 28 percent in 1983/84. Output from Yorkshire went from about 23 percent to 29 percent. The biggest shrinkages were in Scotland and Wales. Each accounted for about 12 percent of output in the forties – 6 to 7 percent in the eighties. The Northumberland and Durham fields of England went from 19 to 12 percent; West Midlands from 14 to 10; the South Midlands from 7 to 8. North Nottingham has done particularly well. For the post-1963/64 period on which data are available, North Nottingham was the only area not to drop output sharply and so raised its output share from 6 to 13 percent.

The West German coal industry operates in four basins of widely different sizes. The largest by far is the Ruhr. The next is the Saar. Then come Aachen and what is now called Ibbenbüren.[5] *Rates* of contraction have been greater for the larger basins. From the postwar output peak in 1956 to 1984, Ruhr output more than halved – from 125 million tonnes to 61 million tonnes. Saar output went from 17 million tonnes to 10 million tonnes; Aachen, from 7 million tonnes to 5 million tonnes, Ibbenbüren, from 2.5 million tonnes to 2.3 million tonnes. (These are respective percent declines of 40, 29, and 11 percent.) Thus, the Ruhr share in output went from 82 to 78 percent; the Saar, from 11 to 13 percent; Aachen, from 5 to 6 percent; Ibbenbüren, from 2 to 3 percent.

CHAPTER 5

Coal in Communist countries

As noted in Chapter 4, the centrally planned or Communist countries constitute the largest and most regularly expanding portion of the world coal market. These industries generally grew steadily both prior to and after World War II. The war, of course, disrupted production. The great bulk of communist coal production occurs in China, the Soviet Union, and Poland.

The other members of the Council for Mutual Economic Assistance (CMEA) have more modest solid fuel output. When converted to coal equivalents, the lignite of East Germany and Czechoslovakia put them a bit ahead of Canada as a solid fuel producer – ranking tenth and eleventh, respectively (recall Table 4.1). East Germany produced about 86 million TCE of solid fuel in 1984; Czechoslovakia, 71 million. Moreover, in both these last cases, lignite comprises a large portion of the total. East German solid fuel output in 1984 consisted entirely of 283 million tonnes of lignite. Czech output includes 26 million tonnes of hard coal and 103 million tonnes of lignite. The other CMEA countries produce even less; 23 million TCE for Romania; 17 million TCE for Bulgaria; 13 million TCE for Hungary.

The history of coal production and use in these countries involves both similarities with and differences from OECD countries. The critical similarities are movements (albeit differing in extent compared to both OECD countries and each other) towards increasing oil and gas use and reacting to higher oil prices since 1974. The special problems are more numerous. The most critical are those of being in a centrally planned economy. The literature on CMEA energy regularly discusses how the energy sector endures the standard problems of communist countries in devising and satisfactorily implementing economic plans. The Soviet Union faces the further difficulties that it is enduring depletion of its traditional supply sources in the European part of the USSR. Discoveries of substantial resources of coal, oil, and natural gas have been made in Siberia.

The Soviets face difficult decisions about how best to resolve the problems of utilising these Siberian resources. Further issues arise in relating these developments to Soviet energy trade with western countries and the CMEA countries. The last relied heavily on the Soviet Union to meet increases in energy demands and particularly those for oil and gas. The USSR, moreover, deliberately moved slowly to adjust oil prices to CMEA countries (and within the USSR) in response to rising world oil prices.

China's energy practice of self-sufficiency is broken to meet pressing needs. China is much less further along the path towards reliance on oil and gas than the other communist states. Oil and gas development has proceeded more slowly than hoped, and coal industry expansion has been critical to meeting energy demands. As with the Soviet Union, the distances between coal producing and industrial areas cause logistical difficulties.

The review here begins with a sketch of the history and prospects of USSR coal production. This is related to trends in Soviet energy consumption. This leads to examination of the role of coal in the different CMEA states. Finally, the Chinese situation is viewed.

Coal in the USSR

The Soviet Union's coal industry is scattered over several different parts of the country. The Donets basin of the Ukraine long has been the largest single producing area (see Table 5.1). It was producing 94 of the 166 million tonnes supplied in 1940. After wartime disruptions, 1950 production was at roughly 1940 levels. A decade later output had almost doubled. Subsequent growth was much more modest and output peaked in 1976 (Hewett 1984, p. 86 and US Congress Office of Technology Assessment (OTA), p. 84). A major problem is that after exploitation dating from the middle nineteenth century, depletion is far advanced. An average depth of 500 meters and seams thinner than a meter prevail (OTA, p. 86).

The western Siberian Kuznetsk basin long has been the second largest producing region, and output growth has continued into the 1970s. Another important mining area is Kazakhstan. While its first principal basin, Karaganda, seems to have had a slowdown in the middle 1970s, the Ekibastuz lignite basin has grown rapidly. Another promising lignite field, Kansk-Achinsk, is being developed in eastern Siberia.

The net impact has been rising coal output from 1950 to the middle 1970s. Output peaked around 724 million physical tonnes (according to OTA). The German Statistik der Kohlenwirtschaft sets 1978 output at 557 million tonnes of hard coal and 163 million of lignite and 1984 output at 555

Table 5.1 *Coal production in the USSR, selected years, 1940–1983*

	1940	1950	1960	1970	1975	1980	1983	Change in tonnage 1975-83
Total	166	260	510	624	701	716	716	15
European USSR	117	171	327	355	366	338	324	-42
Donets	94	95	188	217	223	204	196	-27
Moscow	10	31	43	36	34	25	21	-13
Pechora	0	9	18	21	24	28	28	4
Urals	12	33	59	54	45	44	44	-1
Other Europe	1	3	19	27	40	37	35	-5
Asian USSR	47	89	183	269	335	378	392	57
Ekibastuz	0	0	6	23	46	67	72	26
Karaganda	6	16	26	38	46	48	49	3
Kuznetsk	22	38	84	113	138	144	147	9
Kansk-Achinsk	0	2	9	18	28	35	40	12
South Yatutia	0	0	0	0	0	3	4	4
Other Asia	18	33	58	77	77	81	80	3
Discrepancy	2							

Sources: 1940: Unternehmensverband Ruhrbergbau: *Die Kohlenwirschaft der Welt in Zahlen,*
1961 edition, p.130; other Asia may include regions separately reported by CIA.
Other Years: US Central Intelligence Agency, *USSR Energy Atlas,* p.36.

and 157 million. Thus, it shows a move from 720 million to 712 million tonnes in toto. These levels contrast to 1950 levels of 261 million and 624 million tonnes in 1970 (see Table 5.1).

The Soviet Union's coal industry faces many problems. Standard Soviet difficulties of securing and effectively employing labour and capital prevail. OTA cites estimates that over a million workers were employed in Soviet coal mining in the early 1970s and that subsequently output per worker has grown more slowly than output (p. 94–5). Output per worker was about 650 tonnes per year or, assuming a 250 day year, below 3 tonnes per day, 50 percent more than those cited in Chapter 2 for high cost European industries. The 1979 level was about 840 tonnes per year.

Additional problems arise of transporting Siberian coal to market. The same alternatives considered elsewhere – rail, slurry pipelines, or transmitting electricity from minemouth plants – have been evaluated for the Soviet Union. Organisational problems interact with the intrinsic uncertainties to preclude rapid resolution of the issue. Moreover, all solutions imply that substantial investments are required to make the coal available. Thus, what the electric power sector saves on power plants by not building nuclear plants may be spent on coal transport or electricity transmission investment. Coal use may not be a capital saving alternative, and the

process of providing funds to different ministries might cause a misallo-
cation of resources.

These difficulties are aggravated by the high proportion of lignite in the
Siberian reserves. Lignite at best is expensive to move because more tonnes
are required to produce a given heat content. Siberian lignite is also subject
to spontaneous combustion, deterioration, and being blown from cars in
transit. Thus, building power plants and transmitting the power on high
voltage lines is the preferable alternative. Here, too, severe implementation
questions arise.

Coal output growth has been associated with a declining role of coal in
Soviet energy consumption. Here discussion is limited to the 1980 Soviet
Union situation. Hewett (1984, p. 107) conveniently compares 1970 and
1980 Soviet energy consumption to that in the United States and western
Europe. The Soviet Union proves to have moved considerably away from
heavy reliance on coal. The 1980 coal share was 31 percent compared to 23
percent in OECD Europe and 22 percent in the US. The critical counter-
part of the higher coal share was a lower oil and gas share – 64 percent for
the USSR, 65 percent for OECD Europe and 70 percent for the USA. In
addition, the Soviet nuclear share was 1 percent; a 4 percent share occurred
in the USA and OECD Europe. Hydro shares were 3 percent in the USSR,
4 percent in the USA, and 7 percent in OECD Europe.

A further difference is that the USA and USSR, with large domestic gas
production, had higher gas shares than OECD Europe – 27 percent for the
USA and USSR and 14 percent for OECD Europe. The respective oil
shares are 43 percent, 37 percent, and 51 percent.

These overall figures obscure more pronounced differences in energy use
in individual sectors. In particular, coal shares of about 25 percent prevail
for direct fuel use in the Soviet Union compared to the 18 and 19 percent
shares for coal in industry in USA and OECD Europe and of 1 and 11
percent for other stationary uses. Conversely, a lower coal share prevails in
Soviet electricity compared to the USA or OECD Europe.

Stagnation in coal output occurred simultaneously with a plateauing of
oil output. It has been around 610 million TOE or 870 million TCE from
1980 to 1984 (BP, p. 4). Nuclear power has grown sharply but lags behind
schedule. The Soviet Union in 1984 was fourth behind the USA, France,
and Japan in nuclear development. Its 35.7 million TCE compares to 128
million TCE in the USA, 55 million TCE in France, and 43 million TCE in
Japan (BP, p. 24). The only Soviet energy sector that has performed well is
natural gas. In 1960, gas output at about 54 million TCE was about a
quarter the heat content of oil output. By 1970, gas output was at 233
million TCE (Hewett 1984, p. 31). By 1985, output of 827 million TCE was
97 percent of the heat value of oil (BP, p. 21).

Thus, gas has been the critical source of energy expansion. Some Western commentators give the impression that gas was the almost accidental salvation of an ill-focused energy planning effort. The evidence they present can alternatively be interpreted as indicating that gas is the cheapest to expand fuel. Similarly, the suggestion is made that some combination of accelerated coal and nuclear development, resuming oil output growth, and efforts to encourage lower use of energy is needed to supplement gas. Many writers on Soviet energy seem to view gas as a temporary source of expanded supply. This need not be the case. What does emerge reasonably clearly from the literature is the Soviet Union has many physically attainable energy alternatives, great uncertainty about the relative attractiveness of each, and an administrative structure ill-suited to resolve the situation expeditiously.

The ultimate resolution then is unclear. One can imagine the continuation of gas expansion for several decades. This combined with ample energy supplies in western Europe could slow sales to Europe and increase gas supply to CMEA Europe. Coal output might resume growth or might be displaced by nuclear power. Clearly then, future coal prospects in the Soviet Union are extremely unclear as are the much discussed prospects for energy exports.

The Soviet Union's energy experience is the quintessence of the defects of concentrating solely on the quantity of known resources. Soviet reported energy resources are the largest in the world. However, their development is problem plagued. Part of the difficulty is that the underlying economics are less favourable than a critical viewing of the quantity figures might suggest. The primary barrier is locational. The easiest to produce coal, oil, and gas resources are in Siberia. Thus, difficulties arise in developing these resources because of the underdeveloped state of that region.

Further impediments are caused by the need to transport the fuel long distances to markets. In the coal case, the situation is aggravated by the large amount of resources that are low grade lignites. The cost of shipping useful energy is increased by the low heat content by tonne. In addition, as noted, much wastage occurs. Finally, the Communist system is detrimental to efficient economic development.

Coal in Eastern Europe

As the review above of output patterns suggests, the six CMEA countries differ radically in their coal and energy situation. At one extreme, Poland produces more calories than it consumes. Since the output is predominantly coal, oil and gas are obtained to meet specialised needs and probably to serve some markets where their use is more convenient. Poland thus is both a substantial exporter of coal and an oil and gas importer.

Historical experience and the available literature suggest that Polish coal output *expansion* potential has been and remains greater than in the other nontrivial CMEA producers, East Germany and Czechoslovakia. Depletion problems in Poland's established Silesian fields may be offset by opening mines in the newly discovered fields near Lublin.

It is often stated that Polish coal exports are governed more by the need for hard currency than by the economics. These statements overlook the perennial drawback of Communist economies of their inability to develop rational pricing systems. As a result, the true economics of coal trade are probably unclear to the Polish government and certainly poorly known to others. Statements about the urgency of any country to gain foreign exchange usually arise because of efforts to maintain too high an exchange rate. Thus, exports are more remunerative than domestic currency values at the prevailing exchange rates would indicate.

While Polish output has grown, East Germany and Czechoslovakia have found it preferable to rely on other fuels to meet energy consumption growth. Rumania is a CMEA special case of a quite different nature. Rumania was one of the first places in the world to develop a significant oil industry whose output began declining around 1977. Rumania has turned to reliance on imported energy.

Collectively, CMEA Eastern Europe has remained far more coal dependent than the Soviet Union. However, the variation among countries, reflecting differences in energy supply conditions, is quite considerable. Rumania long has had heavy reliance on oil and gas. The major change has been a rise in the gas share (Table 5.2). Those countries with significant coal resources have maintained high reliance on coal but have increased the proportion of oil and gas used. Finally, Bulgaria and Hungary have moved away from predominant dependence on coal. In Bulgaria, the coal role remains above that in the Soviet Union. In Hungary, coal, oil, and gas are roughly equal in importance.

As noted, oil and gas from the Soviet Union has been a major component of CMEA energy supplies. A recurrent concern in western writing on Communist energy is the future of these sales and particularly the consequences of any cutbacks. These discussions encompass the areas of Soviet potential, energy supply alternatives, and the prospects for adopting energy supply alternatives. For present purposes, only views on coal expansion prospects are germane. The basic points are that coal supply expansion ambitions are modest and the prospects for attaining them are questionable.

An obsolete Office of Technology Assessment (OTA) review of CMEA coal expansion plans indicated that the most ambitious expansion plans

Table 5.2 *CMEA country energy use in selected years, 1950–83 (million tonnes of coal equivalent)*

	Solid	Liquid	Gas	Electricity	Total
		Bulgaria			
1950	3.1	0.3	0.0	0.0	3.4
1960	8.5	1.4	0.0	0.2	10.2
1970	19.4	11.8	0.6	0.3	32.1
1980	21.0	19.6	5.0	1.7	47.3
1982	22.5	19.2	6.1	2.0	49.8
1983	22.5	19.2	6.3	2.2	50.2
		Czechoslovakia			
1950	29.3	1.0	0.0	0.1	30.5
1960	48.5	2.5	1.8	0.3	53.1
1970	61.1	12.7	2.7	0.9	77.4
1980	63.5	22.4	10.2	1.4	97.4
1982	64.3	19.2	9.8	1.4	94.7
1983	65.1	18.8	11.4	1.5	96.9
		East Germany			
1950	47.6	0.3	0.0	0.0	48.0
1960	77.3	1.6	0.1	0.0	79.0
1970	88.8	14.8	0.7	0.3	104.5
1980	86.1	23.6	10.3	1.8	121.8
1982	87.6	22.7	11.2	1.7	123.2
1983	87.8	19.9	12.2	1.9	121.8
		Hungary			
1950	6.7	0.6	0.5	0.0	7.8
1960	15.4	2.5	0.7	0.1	18.7
1970	16.9	7.6	4.1	0.4	29.1
1980	12.8	14.5	12.4	0.9	40.6
1982	13.6	13.2	13.2	1.1	41.1
1983	12.8	12.7	12.9	1.4	39.8
		Poland			
1950	41.0	0.5	0.3	0.1	41.9
1960	70.8	2.8	1.0	0.1	74.7
1970	98.2	10.3	7.6	0.2	116.3
1980	143.2	20.8	12.4	0.4	176.8
1982	131.5	16.5	12.3	0.1	160.4
1983	131.7	17.0	12.7	0.0	161.4
		Rumania			
1950	2.5	4.2	4.3	0.0	10.9
1960	5.5	7.6	13.9	0.0	27.0
1970	14.0	13.8	33.1	0.0	61.0
1980	23.5	23.5	51.4	1.6	100.0
1982	23.0	20.5	54.9	1.6	100.0
1983	25.5	18.6	55.8	1.5	101.4
		CMEA			
1950	130.2	6.8	5.1	0.3	142.5
1960	226.0	18.3	17.5	0.8	262.6
1970	298.5	70.9	48.8	2.1	420.3
1980	350.0	124.3	101.7	7.8	583.9
1982	342.5	111.3	107.5	7.9	569.2
1983	345.5	106.1	111.3	8.6	571.5

Source: United Nations, *Energy Statistics Yearbook* 1982 and 1983.

were in Poland. Plans called for hard coal output of 260 million tonnes by 1990 and raising lignite output from its 37 million tonnes 1980 level to 115 million tonnes in 1990. This was characterised as raising output from 2.30 to 3.75 million barrels of oil per day equivalent. Using conversion factors suggested by BP, this is a move from 164 million TCE to 267 million TCE. A 1983 Economic Commission for Europe report sets the 1990 goals at 192 million TCE and the 2000 objective at 200 million TCE.

According to OTA, East Germany was hoping to raise lignite production to 300 million tonnes. Czechoslovakia expected stationary hard coal output but a rise in lignite production to 109 million tonnes. Rumania was seeking to more than double its modest hard coal and lignite output. According to BP estimates for 1984 (Table 4.1), East Germany with 283 million tonnes of output was well ahead of plans. Poland's growth has been far behind schedule, and the rest of the block has not done well.

OTA noted several problems that hinder the attainment of these plans. These include deficiencies in the quantity and quality of surface mining equipment, labour recruitment difficulties, and possible need to respond to the air pollution impacts of lignite burning and the land disruption of surface mining. OTA estimated that actual output for the block as a whole would range from 4.99 to 5.46 million oil equivalent barrels per day in 1985 compared to a plan of 6.09 million. The respective 1990 numbers are 5.44 to 5.98 million compared to plans of 6.86 million. (In TCE, this is a 355–388 million tonne range compared to a 433 million goal in 1985, 387–425 million in 1990 versus 488 million). The 1984 actual was 409 million TCE.

Coal and energy in China

China remains the most coal dependent major country in the world. It also still has heavy reliance on the wood and waste energy sources. Coal accounted for about 90 percent of the fuel (other than wood and waste) used in 1965, and at least 70 percent in 1983.[1] The coal share dropped from 1965 to 1979. Then oil and gas consumption began to decline (in absolute terms) so the coal share rose.

The International Bank for Reconstruction and Development (The World Bank) data (1985, p. 3) on end use in 1980 show that coal has a slightly lower share in electricity generation than in total energy use. However, the electricity sector is a small part of total energy and coal use. Only 78.2 million TCE of coal was used in 1980 for generation. Larger uses occurred in industry (189 million TCE) and residential and commercial (90 million TCE).

China faces a problem analogous to that of the USSR. The lowest cost Chinese coal resources are located 1,000 kilometers or more from the main markets. The World Bank estimates that in Shanxi Province in the North output can be expanded by 36–44 yuan per tonne compared to 60–70 yuan costs in the Northeast nearer industrial centers. (One US dollar equals 2.8 yuan.) The net is that with respective transportation costs of 25 and 5 yuan, the delivered costs are 61–69 yuan from Shanxi and 65–75 yuan from the Northeast (World Bank 1985, p. 102). Moreover, the Bank's figures are based on the long-run marginal cost of an unspecified level of expansion. The Bank's discussion suggests that more capacity could be added in Shanxi than in the Northeast without depleting the resources mineable at the postulated costs.

As in the Soviet Union, choices must be made between transporting the coal by rail and building mine mouth power plants. Also as in the Soviet Union, problems arise in coordinating among the coal industry, the railroads, and the electric power industry. Moreover, the Chinese have three types of mine management. The central government, provinces, and towns and villages all operate mines. The last type has been the most rapidly growing. Output was under 10 million tonnes in 1965 and reached 170 million in 1983 (World Bank, 1985, p. 91). Thus, about a fifth of output comes from such mines. Provincial mines have kept output at about a quarter of the total. The share of central government production has declined. The World Bank projects a change in this situation in large part due to the attractiveness of large scale centralised mines for incremental output.

The Bank developed two scenarios for the year 2000. (Both report tonnes produced rather than TCE.) One calls for a rise from 715 million tonnes in 1983 to 1,200 million tonnes of physical output, the other for 1,400 million tonnes. In both cases, central mine output grows faster than that of other sectors. Provincial mine output differs little between the cases; village output differs radically. Thus, centralised output goes from 363 million tonnes in 1983 to 750 million tonnes in the lower 2000 case and 800 million tonnes in the higher. The corresponding figures for provincial mines are 182 million tonnes in 1983 to 240 million or 250 million tonnes. For local mines, the estimates are from 170 million tonnes to 210 million or 350 million tonnes (World Bank 1985, p. 97).

However, these forecasts may not adequately consider that the probable persistence of the critical forces stimulating local mines. These local mines have the advantages of lesser requirements for decisions by the central planners. Less capital is needed, and more flexibility exists in decision making. The ability to maintain surface mined output in the eastern United States in the face of alleged depletion problems is a reminder that other influences must be considered.

Finally, China has conducted discussions with western coal producing companies such as Occidental Petroleum and Shell. The discussions involve participation in developing mines in China. Such mines would provide an infusion of advanced technology and a potential to develop a significant export trade. Occidental reached an agreement, but Shell announced it found the prospects unattractive. The World Bank estimates Chinese coal could become competitive in the world coal trade. Again, the prospects are unclear. Given the difficulties involved, the actual emergence of significant Chinese coal exports is not expected until sometime well into the 1990s. With the existence of exports uncertain, forecasts of levels are of dubious value.

Conclusions

While the Communist block is a leading coal producing and consuming region, many problems prevail. All evidence suggests China will remain heavily reliant on coal and the key question is how much output can be expanded. In contrast, the USSR and the CMEA countries have many options. It is unclear that even the countries themselves possess enough information to determine what is preferable. Natural gas may prove far more than a stopgap before a coal-nuclear choice is made.

Notes on the literature on Communist countries

For many years, western observers have regularly provided commentaries on the Communist economies. Increasingly, energy has been a major element of the discussion. Thus, the Joint Economic Committee of the US Congress periodically commissions a series of papers on the subject and various specialists publish on the area. For present purposes, Ed A. Hewett, a regular contributor to the literature, published in 1984 a useful, lucid overview of Soviet energy including discussion of relations of CMEA countries. His references led to other materials including a helpful Office of Technology Assessment study. Another prolific and helpful writer in the area is Jonathan Stern. China has been the subject of useful surveys by the World Bank. Readily available data sources such as the UN, BP, and German coal reports noted in Chapter 2 provide further material.

CHAPTER 6

Coal in the USA – the public policy issues

The essential aspects of the market side of US coal were covered in prior chapters. This chapter concentrates upon the many public policy issues that have arisen. Discussion begins with general economic principles relevant for appraising public policies. Environmental regulation, rules to improve worker health and safety, federal coal leasing, and coal research are then discussed. Given the vast available literature on these subjects, a summary view suffices here.

The theoretic background

Rationales exist for all US coal policies. Serious environmental damages are a burden on the society that should be alleviated. The theory and practice of public finance advocates taxing economic rents – profits in excess of a reasonable return on capital – from mineral exploitation. Modern governments believe that assistance should be provided to the innocent victims of economic change.

The most fully accepted argument for intervention relates to what is variously known as the publicness, collective consumption, or nonexclusivity problem. This is the situation in which it is impossible or at least infeasible or undesirable to exclude anyone from enjoyment of a resource or commodity. Classic examples of inherent inabilities to ration consumption are defense and clean air. Uncrowded parks, roads, and other facilities are examples of where exclusion is possible but undesirable. Given these conditions, members of society can use governments to raise the money to insure that the services are made available.[1]

The degree of crowding is the critical determinant of when a zero price is optimal. Once substantial crowding occurs, access charges are needed to limit use. Thus, incomes are generated. It becomes feasible for private owners profitably to own the resources and control use, and no efficiency case can be made for government ownership. The frequently encountered case for retention of public lands for recreation centers on the need to protect the land from crowding actually makes it desirable to permit private ownership that imposes charges.

In economic theory, market failure is the inability to respond accurately and fully to the demands of present and future consumers, given their actual incomes. This defect is termed (economic) inefficiency. Most who seek political reversals of market decisions claim inefficiency is involved. However, often what has actually occurred is merely that the efficient solution is unsatisfactory to the interested parties. Society can choose to intervene if it believes that, for such reasons as impoverishment or moral superiority, some groups should receive greater consideration.

These arguments usually are abused. Too many people who are neither impoverished nor morally superior benefit from intervention. The measures used to provide assistance almost invariably do so more expensively than necessary. Environmental claims suffer from these drawbacks.

Intervention might increase competition. Various views prevail about the extent of monopoly, the degree to which it is natural, and the ability of governments to effect improvements and no satisfactory way has been devised to resolve the debate. However, as argued in Chapter 3, monopoly prevention is not a significant problem in coal markets.

The discussion here is highly critical of the policies reviewed. While this conflicts with the strong public support such programs receive, numerous economic analyses of these policies have presented similar attacks on the procedures. The principal difference between prior works and this discussion relates to policy objectives. Many other writers accept the policy goals and suggest better ways to attain them. Research and work as a public policy consultant to the US government has led me to question the wisdom of the goals.

US energy regulation

Coal has been a far less favoured fuel in the United States than in Europe. Few promotional policies have been undertaken. The critical policy impacts have been those of decisions made for concerns other than the role of coal in energy consumption. Through the middle 1960s, the coal industry was the incidental beneficiary of protection accorded the US oil industry under various oil-import-control programs. Nevertheless, coal use declined sharply (see Chapter 3).

Since the middle 60s, US policy has imposed numerous impediments to coal development. First and most clearly justifiably, the oil-import program was changed in a fashion particularly unfavorable to coal. In 1966, quotas on the import to the East Coast of the heavy fuel oil that is the principal competition for coal were increased to levels greatly exceeding prevailing demands at world oil prices. The result was a massive shift of East-Coast utilities from coal to oil.

Moreover, much of the response produced barriers to reconversion to coal. Many utilities stopped building plants with the capabilities to burn coal, and facilities to receive and use coal were removed from some old plants and the land was dedicated to other facilities.

Subsequently, the critical policy changes have increased the cost of producing and using coal. The first major new programs were environmental. The 1969 National Environmental Policy Act (NEPA) established procedures to inspire greater concern for the environmental effects of federal government actions. Court cases turned the Act's requirements for studies of the environmental impacts of major federal actions into a complex evaluation process. Stringent standards were adopted such that a wide variety of actions were defined as major ones affecting the environment and many impacts and alternative approaches had to be considered. For example, setting a basic leasing policy, holding a lease auction, and letting a mine begin operation are separate major actions.

All alternative energy forms had to be considered in actions involving any specific energy type. Examination of the process suggests that its main impact was the imposition of long delays in decision making. Intervenors often succeeded in getting courts to rule that the evaluations were inadequate. The decision-makers moved to undertaking extensive, protracted studies to disarm critics. However, court cases continued. Even elaborate studies could be considered defective. An effective way was found to delay actions. Projects were more likely to be killed by these delays than by reviews. Ultimately, more projects proved able to secure legally acceptable environmental appraisals.

Major, more direct impacts were caused by imposition of stricter air pollution regulation and institution of surface mine reclamation laws.

Air pollution laws have been tightened as concerns increased. Sulfur dioxide regulation has been the most critical. The rules impose expensive restrictions on coal use. Regulations propounded in 1971, among other things, severely limited the amount of sulfur dioxide that could be emitted from new large-scale fuel-burning facilities. This required alteration of behavior by US utilities. Many had previously relied on eastern coals that when burned produced more emission than the new standards allowed. Thus, buying such coals for new plants was not possible without changing operating practice.

Three alternatives exist – finding a naturally lower sulfur fuel removing the sulfur before combustion, or employing devices to remove the pollutants after combustion but before discharge to the atmosphere. Suggestions were widespread in the early 1970s that scrubbers, devices for removing sulfur oxides after combustion, had been perfected and were a cheap, reliable control option. Thus, policy makers and some in the coal industry asserted that compliance would occur quickly with the rapid installation of scrubbers. In fact, scrubbers proved harder to develop, more expensive, and less reliable than optimists proclaimed.

The results were some shifts to oil by eastern electric utilities, shifts to low-sulfur western coal by many utilities in the East-North-Central states and Minnesota, and a growing conviction among utilities that nuclear power was the preferable long-run alternative. In response to the shift to western coal, what has been frequently described (e.g., by Ackerman and Hassler 1981) as an unholy alliance of environmentalists, eastern coal interests, and the union representing eastern workers inspired in 1977 major changes in air-pollution law.

Among the numerous provisions was the requirement that pollution abatement involve the use of best-available control technology (BACT). The intent was to move the compliance choice more towards scrubbers and less towards western coal. In practice, problems arose in interpreting the law.

Scrubbers cannot remove all the sulfur; the residual pollution from use of a very high-sulfur coal could exceed the emissions from unscrubbed use of a low-sulfur coal. At least for this reason, the sulfur content of the fuel had to be considered in setting rules. This created the potential for (cost minimising) variation in rules that fully credited users of low sulfur coal for the benefits of the initial low sulfur content. The actual policy set the credit at less liberal levels.

Other aspects of the revisions included a provision that allowed prohibition of the use of nonlocal coal in existing plants if the local economy was damaged. Since damage is difficult to establish, the closest thing to application was loosening of air-pollution standards in Ohio to stave off application of the local-coal provision.

Still other clauses divided the country into two parts – nonattainment areas in which air pollution standards were not met and prevention of significant deterioration (PSD) areas where they were. In nonattainment areas, expansion of industry was severely limited by requirements that no new facility that produced pollution could be opened unless an offsetting reduction in pollution occurred elsewhere.

In PSD areas, limits were set to the allowable *increases* in pollution. The law, moreover, established three levels of allowable increase – the most limited applicable to areas such as National Parks where air quality preservation was considered particularly critical. More liberal increases were allowed where the concern was less severe.

Most criticisms of the rules have stressed the role of PSD in preventing extensive moves of industry from industrialised areas and downplayed the pressures on such areas of the nonattainment rules. It seems more accurate to describe the policy as a general discouragement of economic expansion.

A vast economic literature has arisen suggesting that environmental regulations seek to attain their goals in inordinately expensive, time consuming rules. The evidence also suggests that policy makers have set their objectives on the basis of imperfect estimates of both damages (particularly to health) and what causes them.

The critiques may be too timid. The stress is on the excessive cost of meeting accepted goals. The widsom of the goals had received less challenge. Evidence exists, for example, that the Lave and Seskin estimates of the health effects of air pollution used to justify present policy were far too high (see Ramsay 1979). Nevertheless, they are still widely cited in policy debates. The Office of Technology Assessment used them in 1984 to justify action on acid deposition. Damages to lakes are too small to justify massive control outlays so another rationale is needed. Too much abatement may be sought. Moreover, given that several pollutants occur together, the estimates cannot determine which one or combination is the culprit, and the wrong problems may be emphasised (see Wilson *et al.* 1980).

A 1969 mining disaster inspired Congress severely to tighten coal-mine safety regulation and introduce controls of impacts on worker health by imposing elaborate, highly detailed rules. Here too questions exist about the wisdom of the action. The approach probably was inferior to reliance on collective bargaining. The effects of the law were difficult to determine. The work rules should have lowered output per worker. However, other forces such as an influx of inexperienced workers also influenced coal industry productivity. Thus, while output per man day declined in the 1970s and rose again in the 1980s, the contributions to the decline of different influences could not be determined satisfactorily from the available data.

US coal leasing has had a turbulent, difficult to summarise history. A steady stable leasing program could not be maintained. A moratorium on leasing was instituted in 1971 and lasted until 1981. By 1983, leasing was halted again by Congressional complaints about administration, an elaborate investigation, and Department of Interior efforts to meet the criticism. Resumption is unlikely until at least 1987. The history of these moratoriums is too complex to treat fully here. Developing a policy that satisfied NEPA requirements was the main problem in the 1971–81 period. The second moratorium arose from Congressional fears that the administration of leasing did not adequately satisfy the legal requirements. Throughout the 1971–86 period, two distinct issues have been involved – environmental impacts and whether the government properly charged for the right to lease. The importance of the two concerns has shifted over time.

The moratorium initially was imposed in 1971 using the administrative discretion of the Secretary of the Interior. The long delay in reinstituting leasing was due mainly to difficulties in proving compliance with the environmental regulations arising out of NEPA. However, in the middle of the effort to formulate a new coal leasing policy, Congress chose in 1976 to revise totally the law governing coal leasing. Deciding how to administer these new provisions became a concern for Interior.

Through historical accident, US coal-resource ownership is split between the public and private sectors. In the traditional coal producing regions east of the Mississippi, private ownership predominates. About 60 percent of the coal west of the Mississippi and about 72 percent in the six states (Colorado, Montana, New Mexico, North Dakota, Utah, and Wyoming) with both major coal deposits and heavy federal ownership is federally owned (US Department of Interior, 1979, p. 2–5).

Leasing proceeded modestly until the late sixties, when it spurted. Questions immediately were raised about why so much leasing occurred when output was far below historic peaks. The leasing was in anticipation of the rapid rise of western coal output. The prospects, which were realised, were evident in the late sixties. The fears, nevertheless, necessitated the leasing moratorium.

Public land policy long had led to fragmentation of property rights on federal coal-bearing lands. Ownership of the subsurface is separated from that of the surface and the subsurface continues to belong to the federal government. An indivisible property is treated as if it were two; this guarantees conflict.

The subsurface cannot be exploited without affecting the rights of the surface-property owner. While a conceptually acceptable basis for compensation can be defined, its implementation can be difficult and also be considered inequitable. Compensation could be limited to the actual costs to the surface owner. No legal system can guarantee exactly this outcome. A system based on damage valuations may be biased; where the government is simultaneously the conflicting claimant and the arbitrator, the surface owner may fear undercompensation.

However, if the surface land owners have priorities in demanding compensation, they often will be paid more than the damages to their property. If the surface owners are paid before the government, they can and will seek to extract all the gross profits and not just those equal to their costs (see below).

Further difficulties are caused by the grants of land to railroads, past leasing of coal, and past accords with surface owners. All lead to further diminution of property rights. To promote railroad construction, grants of land were given those completing lines. The area had previously been divided into square sections and railroads were awarded every other section along the right of ways. Because of the appearance on the maps of these allocations, the process was called the checkerboarding of the West. Over the years, moreover, land was dedicated to nonmining use, often with transfer of the surface to private ownership, coal was leased, and accords were reached with surface owners.

Thus, much coal was already controlled. Additionally, much of the coal was most profitably used as an extension of existing leases. In other cases, railroad properties had to be combined with federal properties to allow an efficient operation. The result was that often the parcels available for leasing could most profitably be used by those with existing rights to adjacent coal resources.

Coal leasing legislation in the 1970s

In 1976, Congress passed, by overriding President Ford's veto, the Coal Leasing Amendments Act. It radically altered coal leasing by eliminating the long-standing ability to secure preference right leases, ones granted without either competition or charges if a discovery of coal had been made. Restrictions were imposed on the ability to explore. The results had to be given (on a confidential basis) to the Department of the Interior. All coal had to be leased under competitive bids with a requirement that surface mines also be subject to the 12.5 percent minimum royalty rate long applied to oil and gas leases. Restrictions were imposed on the size and duration of coal holdings. To offset the disincentives to exploration, the Department of Interior was supposed to establish its own exploration program.

To win political support, the state share in gross federal receipts from coal was raised to 50 percent. Comprehensive land use planning was supposed to precede leasing to prevent harm to other uses.

The bidding was to be guided by the fair-market-value principle widely employed in US government decision making. However, severe practical difficulties arise in implementing the concept. Concerns involve (1) whether the market is adequately competitive, (2) the comprehensiveness of information about what has occurred in the market, and (3) what are appropriate surrogate value estimates, when markets are noncompetitive or data are not available. These difficulties inspire nearly impossible-to-settle arguments about the adequacy of payment to the US government for coal (and other mineral) leases and other property sales.

Worry about the failure to develop leased resources led to increasing radically the restrictions on lease holding. Previously, the only significant prohibition was a 1920 ban on railroad holding of federal leases. Other leaseholders were supposed to be diligent in developing the property. The Department of the Interior presumed that, as profit-making institutions, leaseholders would operate when remunerative. Thus, every private decision automatically was diligent in the socially relevant sense of making the resource available when it was profitable to society to do so. (Note that if anything, the private payoff to development is greater than the social. The latter may be lower because of environmental degradation.)

The Amendments reject the view and define diligence as coming into operation within a decade of leasing. In addition, any company that had held and not developed any coal lease for more than a decade was banned from bidding on any further federal lease for any type of mineral subject to leasing. In addition, limits were set on the allowable size of individual leases and on how much acreage could be held by any company in a given state or nationally.

A further change in policy was imposed by the 1977 Surface Mine Control and Reclamation Act. The Act was concerned with all the surface impacts of coal mining no matter who owned the coal or the surface. A complex set of reclamation rules was imposed. Again, the details are too particular to merit inclusion here. However, one provision required that those seeking to lease federal coal secure the consent of "qualified" surface owners (essentially those who lived on, farmed, or ranched the land). This provision created potential for transfer of significant income from the federal leasing program to the surface owners. In fact, the provision should have virtually eliminated lease bonuses. The permission of the surface owner had to proceed before leasing. Vigorous competition for permission could bid up the price to the residual value of the property after royalties and state taxes had been paid. This would leave nothing for lease bonuses.

The effect of policy

As should be expected, all these policies lowered the attractiveness of coal as a fuel. It is remarkable that the coal industry has done so well given these adversities. Even though the industry has managed to lessen the impacts, the policies should be reformed. In fact, policy reform may be far the most critical need in the coal realm.

In part, the need for reform has been obscured by offsetting developments. World oil prices rose massively. Nuclear power in the United States vanished as a satisfactory alternative. Bitter debate exists on the causes. Electric-utility executives insist that in a world of rational regulations, nuclear power would be preferable to coal in most of the country. The irrationality of regulation makes new nuclear projects undesirable.

Critics of nuclear power assert that the problem is excessive costs. Costs are uncompetitive. The question is the extent to which cost rise was due to regulation, mismanagement, or market developments. Whatever the inherent economics, the regulatory barriers to nuclear power in the United States have guaranteed that no more nuclear plants will be ordered in the foreseeable future (see Gordon 1982).

The coal industry, however, may have gained less from these offsets than it has lost from the impact of higher energy costs on the growth of energy consumption. To what extent this effect is desirable also depends upon how much of the cost increases were attributable to sound policy and market developments beyond the control of the US government and how much was due to unsound policies.

Research and development policies

Efforts to assist development of new coal-using technologies have been plagued with difficulties. In the early 1970s, the impression was being created by at least some of the more flamboyant and thus more publicly known advocates that several alternative technologies for coal utilisation would soon be economically available.

Politicians, particularly in the Carter administration, used these ideas to create ill-advised energy research and development programs, particularly the Synthetic Fuels Corporation, as a substitute for policy reform. The refusal to remove a morass of oil and gas price controls was rationalised by claims that only new technologies would solve the problem. The resulting reactions caused drastic deterioration of the climate for energy research.

This, of course, denotes neither a failure to effect technological advances nor the undesirability of careful efforts to formulate a better research program. Interesting work has been done on such areas as gasification and fluidised bed combustion. The research agenda should not suffer from the defects already noted of past efforts. We should not repeat the discredited notion that we must use coal for everything.

CHAPTER 7

The state and coal – the Western European case

Almost continuously since World War I, Western European coal has been a troubled industry. That war terminated a century of coal industry expansion throughout Western Europe. The war itself disrupted the coal industries. Even when combat did not directly affect the industry, strains emerged.

After World War I, efforts were made to rebuild the industry. Generally, prior output levels were restored. However, Britain faced severe problems that precluded return to prior production levels. Long festering labour–management problems aggravated the British situation. This led, in particular, to a long shutdown of the industry in 1926. The dispute was exceedingly bitter, and the two sides could not even agree whether it was a strike or a lockout. The Great Depression and World War II caused further setbacks.

After World War II, those concerned with coal argued that, with the proper effort, the coal industry would make a growing contribution to energy supply. Steps were taken to restructure the industry to ensure the desired results. Britain and France undertook the most formal efforts. Both countries resolved long-standing debates about how best to reorganise the industry by electing governments favoring nationalisation. British coal was placed under a National Coal Board. Charbonnages de France was created to run the French industry.

In contrast, pressures by the occupying powers led to increased fragmentation of the West German coal industry, mainly as a by-product of the breaking up of the amalgamation of steel firms into Vereinigte Stahlwerke. The successor steel firms received coal mines, and a separate company was set up to operate mines independently of the steel industry (see Lister 1960). This process was ultimately reversed.

State involvement in coal had previously existed in Germany and the Netherlands. A state-owned company was set up early in the twentieth century to develop Dutch mines. Over an extended period, the German government acquired ownership of several coal mines and mine-owning steel companies. In some cases, the ownership was partial, and stock was sold to the public.

In addition, the European states own the railroads and thus are a major influence on coal transportation. Similarly, a major consumer – electric power – is heavily government owned. The main German utilities long have had federal, state, or local governments as principal or at least dominant stockholders. The largest German electric utility, Rheinisch–Westfälisches Elektrizitätswerk (RWE), owns the largest lignite mining company. Britain and France nationalised electric power after World War II.

This extensive government involvement in critical parts of the coal market provides greater ability to provide and conceal assistance to the coal industry. Special deals with the electric utilities have prevailed throughout Western Europe. While railrate setting is always difficult to appraise, subsidised rail rates are not used to promote coal sales.

In the early 1950s, visions of a critical role for coal and steel inspired French advocates of political unification of Western Europe to propose a European Coal and Steel Community (ECSC) as a first step. It was to promote free trade and coordinated development in coal and steel. Six countries – Western Germany, France, Italy, Belgium, the Netherlands, and Luxembourg – agreed to join, and the Community started operating in 1953.

In 1957, the six formed two more Communities – the Economic and the Atomic Energy. Britain, Denmark, and Ireland joined in 1973, Greece, in 1981, and Spain and Portugal in 1986. Moreover, in 1967, the executives of the three Communities merged, and ECSC ceased to exist as a distinct *operating* body. ECSC and the two newer Communities survive as separate legal entities, created by separate treaties and one executive manages all three.

From the onset, ECSC had to deal with unresolved coal industry problems. The most severe was that the Belgian industry had a high cost component that would be difficult to integrate into the Community. Failure to resolve this problem proved a precursor of more severe difficulties.

A radical change in the outlook on energy occurred in 1958. The rapid recovery of the Middle Eastern oil industry from the effects of the 1956 Suez Canal closing forced recognition that oil competition would be an enduring challenge to Western European coal. Western European policy analysts also realised that many other alternatives such as US coal were available. Neither the anticipated rises in the prices of these rival fuels nor the expected cost reductions in European coal had emerged.

The legacy of past commitments precluded rapidly responding by cutting coal output. Coal advocates continued to rationalise this protection of workers by contentions that the need to use European coal had only been delayed. Programs were established to effect a gradual contraction of output. Commitments to limit coal output reduction continually are repudiated as the promised recoveries failed to materialise. Rising oil prices in the 1970s inspired proposals to reverse the contraction. European coal mining costs proved so high compared to those of alternatives that these plans were largely abandoned. Many users found oil was still cheaper. Others could use nuclear power, imported coal, and natural gas. Coal policy has involved slowly and expensively realising the drawbacks of protection.

A perspective on coal policy

While often rationalised as justified by other grounds, subvention of coal was designed primarily to treat perceived problems of finding new jobs in the face of alleged unwillingness of workers to move out or new industries to move in. The programs attained wide support from outside the coal industry. Uncritical acceptance of questionable premises is always a problem in economic policy debates. Economic analysis stresses that protectionist arguments arise from parochial (although well-intended) visions that overrate the desirability of the industry and exaggerate the difficulties of adjustment. Thus, assistance programs often deal unsatisfactorily with whatever real problems exist.

Protection of employment (or aiding the unemployed) involves the danger of discouraging acceptance of attractive alternatives when they arise. The efforts can and, in the coal case, clearly have actually attracted into the coal industry uncommitted resources that could and should have gone elsewhere. Aside from massive investments and local recruitment, workers were sought from abroad. Intervention produces various barriers beyond the intrinsic disincentives to relocation to attracting new business. They fear being taxed to pay for the subsidies. Some German critics of the programs contend policy makers actively hindered the entry of new enterprises. Coal intervention thus seems another example of excessive intervention whose usefulness is long past.

The prime rationalisation of aid is the belief that Western European coal is still the only safe energy option, and this still influences energy policy. Many Europeans are too impressed with the vanished era in which domestic coal seemed the only dependable energy resource.

Security of supply policy is best defined as an effort that minimises the (present value of the) cumulative social costs of securing energy. This requires far more than covering needs when the consuming country is at war. Concern must be given disruption of foreign supplies due to wars elsewhere, social upheaval, or strikes in supplying countries. Attention must also be given to protection against domestic-supply disruptions.

European coal was dependable only when continental warfare cut off alternatives. Coal supply was considerably less reliable in peacetime. Strike disruption of coal production occurs more often than cutoff of oil imports.

Wars similar to World Wars I and II seem unlikely. Many supply alternatives exist. None is clearly more prone to disruption than European (and particularly British) coal output. Moreover, the diversity means that loss of any one supplier can be made up by relying more on others. As Eileen Marshall and Colin Robinson (1984) have pointed out, dedication to coal in Britain may lessen security by increasing the damages a worker strike can inflict and encouraging workers to be more militant.

The critical defect of the security of supply defense of European protectionism is that the coal can readily be replaced at a significant cost saving by equally secure coal from the United States, Canada, Australia, and Colombia or by Canadian, US, or Australian uranium.

Existence of many active producers around the world (even if they do not supply Europe) suffices to ensure security. Many energy-producing industries have far lower costs than the European coal industry. These other industries have shown considerable ability to adjust output and distribution to respond to demand changes. Modern transportation technology facilitates the movement of fuels among countries.

The best strategy is to rely almost entirely upon these capabilities to offset disruptions. These responses can be allowed to occur after a crisis develops. This approach will cause temporary increases in energy prices until adjustment is made. Programs of preparatory measures to offset crises probably would cost more than enduring a few periods of price surges.

Some would combine the employment problem and belief in the need to maintain a coal industry into a contention the programs are investments in preserving the skills needed for the industry. This argument exaggerates the benefits and understates the costs of assistance. That a coal industry ever will be needed is questionable, and it is doubtful that the need will emerge soon enough to produce a high enough payoff or a low enough cost to make the investment profitable. The payoff of preservation of skills if and when the industry recovers is, moreover, only the reduced training costs; these may be far smaller than apologists assert. Finally, the bulk of the benefits to maintaining skills will accrue to mine owners. Therefore, the publicness argument for intervention does not apply.

To the extent dishonest rationales were provided, they perpetuated the implicit contract to protect workers. Initial errors were maintained far too long. Clearly, inadequate candor has prevailed. However, illusions about the prospects for European coal also persist. The result has been another expensive ill-advised exercise in intervention.

An overview of protection

Although five of the six original members of ECSC had coal industries, profound differences existed in the original state and subsequent evolution of these industries. The Germans strove to restore the old position; others were wary of excessive German market power emerging. The French had an industry it hoped to expand. The Belgians had to cope with a weak position. The Dutch had a small industry that seemed able to hold its own. The Italian hard coal industry consisted of limited output on the island of Sardinia. Britain joined after coal problems were evident and had already begun its adjustment program. Spain just joined in 1986.

Only four of these countries are worthy and capable of extensive treatment. According to IEA coal reports (1982 and 1984), the Spanish industry consists of numerous small high cost ventures. (An informed observer suggests that direct mining costs can be as much as $120 per tonne.) Spanish problems have received far less discussion than those of longer-term members of ECSC. Two of these, Italy and the Netherlands, abandoned their modest hard coal industries. This leaves Britain, Germany, France, and Belgium as the areas of interest. All have intensive programs to protect the coal industry.

All the logically possible aid types seem to have been employed. Coal imports have been subjected to formal or informal controls. Customers, particularly electric utilities, were pressured into various commitments to buy domestic coal at prices above world market prices. Many explicit and implicit subsidies were instituted. Aid frankly earmarked as subsidies was supplemented by government investment aid. At the very least, this involved below market rates of interest. In addition, periodically debts were cancelled.

The critical distinctions about aid relate to objectives. Three basic goals prevail – assisting current output, relieving the industry of the obligations incurred in past operations, and easing the effects of mine closings. Assistance in the first area included funds to finance investments and cover deficits. The past obligations consisted mostly of commitments to worker pensions. The costs of controlling subsidence and water pollution from closed mines also were subsidised. Direct financial aid was given to workers, retraining programs, and subvention of new industries to employ the workers.

The data suffer from the chronic problem of the inappropriateness of conventional accounting figures for economic analysis. Accounting data simultaneously overstate some aspects of economic costs while ignoring others. Accounting is entirely backward looking. Accounting data report either current outlays or rough efforts to charge current operations for past outlays. The implicit commitments for future outlays are considered too speculative to include.

A further question is the opportunity cost of resources. Defenders of aid to coal assert that the alternative to coal employment is unemployment. Apologists for coal protection also contend that most of the apparent costs are in one way or another unavoidable charges for past actions. This applies to cash outlays on pensions and control of the environmental impacts of closed mines. Similarly, the cost of mine capital and the interest on it has been incurred, and closing will not reduce the loss.

Continuation of output, however, creates commitments to invest more, pay more pensions, and incur more expense to offset environmental impacts. Protectionist programs have prevailed for so long that much, if not most, of the presently unavoidable costs could have been escaped had contraction proceeded more rapidly.

Another less critical consideration is the role of the European Communities. The ECSC Treaty supposedly gave the Communities control over national coal policies and the power to aid adjustment. The divergent interests of the members precluded a strong Community stand of any sort on coal policy. The countries resist Community advice on policy reform and agreed largely to act independently. Some assistance, particularly to displaced workers, was channeled through the Community. A 1964 accord formalised the acceptance of national measures. Otherwise, control was essentially pro forma.

One aid program, that of subsidising the difference between the cost of domestic and imported coking coal, was formulated in 1967 as a Community program. While some Community funds were involved, the bulk of the money came directly from national treasuries. These and other subsidies are nominally subject to Community review under guidelines set in 1965. The review consists of an annual report on outlays. While this is a valuable data source, the report writers invariably state that the subsidies comply with the provisions of the ECSC Treaty, which knowledgeable observers feel clearly is not the case.

Further influences on coal have included additional reorganisations of the coal industry, deals with electric utilities in different countries, and evolution of the steel industries.

The most constrained contraction occurred in the two countries, Britain and West Germany, with the largest coal industries. Belgium made tentative steps toward contraction in the 1950s, established a more formal, more drastic shrinkage program in the 1960s, and moved ahead slowly.

France started output reduction later but has moved more rapidly and thoroughly towards substantial closings.

In comparing the British and German records, a striking disparity between rhetoric and results emerges. In particular, British output has contracted much more than German. The difference in output thus has narrowed substantially. The output trends and the polices that produced them suggest that it is the Germans who have been the strongest devotees of a welfare-state approach to coal.

Coal in the UK

The UK has developed its reaction to coal problems largely employing the institutional framework established after World War II. The only critical subsequent organisational changes were the denationalisation and renationalisation of the steel industry. Even this was less influential than the decisions in the late 1970s of the nationalised steel industry to contract drastically.

The National Coal Board (NCB) has periodically restructured itself, but this has not been a major influence on output. What has been critical about the Coal Board has been its vigorous effort up to the early 1980s to generate support for expansion. Not only did the Coal Board's own publication long express this view, but its Chairman greatly influenced the World Coal study work on Britain and the IEA's coal advisory board. The Thatcher government is much less enthusiastic about coal.

The Central Electricity Generating Board (CEGB), however, displayed considerable resistance to being committed to high-cost fuel sources. In the 1960s, it began to build oil-fired capacity and instituted an extensive program of nuclear development. As a result, considerable restraint was imposed on the growth of coal generation.

The CEGB and the South of Scotland Electricity Board are committed to a heavily nuclear future. CEGB's decision to change nuclear technologies has inspired an inquiry that leaves the nuclear prospects uncertain. Dissatisfaction with the gas-cooled reaction approach used in Britain led CEGB to suggest shifting to pressurized water technology, developed in the US but widely used in Western Europe and Japan. The government set up a study that held 340 days of hearings from 1983 to 1985 that raised numerous issues including the possibility of going back to coal. Those I interviewed in Britain generally believe that blocking the change of technology, slowing expansion, and shifting to imported electricity are more likely than urging new coal plants. Thus, electric-power coal use growth in England and Wales is expected to be limited. (The results of the inquiry have not been released as of early 1986.)

The principal aid to NCB is a government grant to cover its losses. The CEGB and NCB have devised a buying arrangement that attempts to ensure that coal costs no more than would imports to existing plants. However, most of these plants were built on the premise that British coal was the preferable option. Freight rate savings averaging somewhere between £7 and £8 per tonne accrue to purchase of domestic coal by these inland plants, and CEGB agrees to pay a price that includes the value of the freight rate saving associated with supply to inland plants (which undertake about 85 percent of the coal use). The contract also bases escalation on world coal price trends instead of either the cost of living or the cost of British coal mining.

Another development has been the modest weakening of barriers to imports. CEGB is allowed to import about 5 percent of its needs but had not exercised this right up to 1984. (Less than one million of the 85 million tonnes received in 1983 were imports, and despite the British coal miners strike, only 52,000 tonnes of imports were received in 1984.) Contraction of coking coal output has outstripped steel industry contraction so that about a third of coking coal needs are imported. Modest imports by others are also being allowed.

As discussed further in Chapter 9, the effect of these demand developments and efforts to rationalise the coal industry are expected to result in major restructuring of the coal industry but not necessarily a substantial fall in output.

Britain has always been viewed as a lower-cost producer, and my interviews elicited belief that good prospects exist that a renovated industry could meet a large portion of national coal demands. However, these demands are likely to be somewhat below levels before the 1984–85 strike. Those visited see declining coal use in electricity and steel and little prospects for gain elsewhere.

On the supply side, three large new mines are in various stages of development. One is well advanced; the third was formally announced in 1985. Efforts are being made to increase productivity significantly using NCB-developed technologies in the best mines now in operation. These new and refurbished mines could make major contributions to output. In addition, lessening of NCB control over independent surface mining operations has produced rises in such output. Coal imports are now being allowed to be maintained at a higher but still modest level.

This combination of lower consumption and new supply sources implies major pressures on the less productive mines. Another development has been the rise of interest in returning the coal industry to private management. The 1984–85 strike demonstrated clearly that government ownership did not solve labour relations problems. Critics of the Coal Board argue that it displays the inefficiencies associated with government guarantees of aid. According to this view, private operators would be better at lowering cost and ensuring the survival of a British coal industry.

German coal

German preservation of coal has involved creation of many support mechanisms. Most key actors in the German coal industry are involved in several, conflicting aspects of the market.

The German coal industry has had a complex structure that has undergone numerous changes.[1] The initial postwar reorganisations were followed by a variety of readjustments. In the Ruhr coal industry in the 1960s, participants included German privately owned steel groups such as Thyssen, Mannesmann, Hoesch, and Krupp; federal-government owned steel companies such as Salzgitter, and foreign steel interests such as Arbed of Luxembourg and Sidéchar of France. The biggest single entity was Gelsenkirchener – the operator of the former Vereinigte Stahlwerke mines kept separate from successor steel companies. Another major participant was VEBA, the heir to long-time government holdings (at the time the federal government held 42 percent of VEBA but later reduced its holdings to 23 percent). VEBA also controls several electric utilities.

Production in the Saar is by Saarbergwerke – 74 percent of which was owned by the federal government, the rest by the Saar state. Arbed's German firm, Eschweiler, is the dominant coal producer in Aachen and owned capacity in the Ruhr. The rest of Aachen output is controlled by an independent company – Sophia-Jacoba. Preussag, a diversified mining and metals processing company of which VEBA once held 25.5 percent, is the sole coal producer in the Ibbenbüren region.

In 1968 the German government proposed consolidating Ruhr production in a single firm – Ruhrkohle. This decision reflected recognition that past concerns over excessive concentration had lost their validity. Consolidation seemed desirable. Both Eschweiler and the mine of the chemical firm, BASF, chose to stay independent.

Ownership in Ruhrkohle was distributed among the owners of the mines absorbed. As of 1984 about 63 percent was shared among eight steel companies; the individual company shares ranged from 4.7 to 12.7 percent. A 1985 reorganization eliminated the participation of four steel companies (Mannesmann, Krupp, Harpener, and Klöckner). Thyssen retained a 12.7 percent share; Sidéchar, 8.3; Hoesch, 7.9; VEBA, long the largest stockholder since its acquisition of Gelsenkirchener, increased its share from 27.2 percent to 39.2 percent. Vereinigite Elektrizitätswerke Westfalen (VEW), another major utility, previously holding only 0.2 percent, raised its participation to 21.9. Subsidiaries of Ruhrkole hold the other ten percent.

The dominant lignite producer, as noted above, long has been a subsidiary of RWE. It has always produced most of lignite output in the Rhineland. With the virtual disappearance of Bavarian lignite production (once dominated by Bayernwerk, the chief Bavarian electric utility), the Rhine has increased in importance. RWE accounted for 85 percent of 1985 lignite output. More critically, the absolute amounts have grown from 83 million tonnes in 1967 to 114 million in 1985. A VEBA subsidiary long has been the principal lignite producer in Hesse and has become the sole producer in Lower Saxony. VEBA's 1985 share was almost 15 percent.

Many other examples arise of companies being involved on several sides of the market. Coal companies and the utilities participate in coal import trading. Ruhrkohle and Saarbergwerke have invested in coal mines abroad, presumably to enable participation in import opportunities that will result from further reduction in coal output.

VEBA is perhaps the most complex case as a leading coal buyer through its utility subsidiaries, the chief stockholder of Ruhrkohle, a lignite producer, a major coal trader, owner of coal abroad (specifically a 25 percent share of Westmoreland Coal in the US), and a significant participant in petroleum refining and marketing. Ruhrkohle itself has conflicts as the owner of a major electric utility, a participant in international coal ventures, and a stockholder in a major gas distributor, Ruhrgas.

All these complexities are reconciled by extensive government intervention. The coking coal side is the more straightforward. General subsidies are supplemented by subvention of coking coal sales. Here, as elsewhere, import controls act as a further incentive to use German coal. Essentially all German coking coal needs are met by German coal and about 12 million tonnes of coking coal and coke were exported to other Community countries in 1984. France received about 4 million tonnes; Italy and Luxembourg, around 2 million tonnes each, and flows to Belgium, Britain and the Netherlands all were a million tonnes each (see EEC, *Coal*, May 1985).

The portion of German subsidies long granted to exports of coking coal has become controversial. These involve much higher per tonne subsidies than those on domestic sales. Typically, the delivered prices of imports that must be met are equal or below delivered prices to German steel mills. However, transportation distances and therefore costs typically are higher than on domestic sales. Thus, the net revenues (the difference between delivered price and transportation costs) are lower. Lower revenues mean higher subsidies.

As a result, the German government is seeking to eliminate the export subsidies. Some observers, including those in the European Communities concerned about the burden of subsidies, strongly endorse the measures. Some German fears are expressed over the claimed lessened contribution to energy security in other Community countries caused by eliminating the subsidies.

However, this argument only highlights the perversity of stressing protection of the coking markets for coal. The alternative suppliers are industrial nations with whom Western Europe has close, friendly ties. Of the 25 million tonnes imported by the EEC as a whole from non-Community countries for coking ovens, 15 million came from the USA and 6.7 million from Australia. A more significant but presumably transitory problem is that some customers lack coke ovens and rely on German pithead coke plants to process the coal.

An even more complex system is used to regulate electric utility coal use. Pressures have been exerted since 1965 when, under the first law on electricity coal use, tax reductions were given to utilities that agreed to use German coal. A second force is the kohlepfennig, a tax on electricity sales instituted in 1974. The tax started out at 3.24 percent of sales and went to 4.5 percent in April 1976. The all time high of 6.2 percent prevailed from the first nine months of 1979. Then a return to 4.5 percent lasted until the end of 1981. The 1982 rate was 4.2 percent. Subsequently a 3.5 percent charge has prevailed.[2] This tax is an effort to spread some of the cost of subsidising the use of German coal evenly among electricity consumers.

The proceeds historically have been used in part to cover the extra investment costs of building coal-fired rather than oil-fired power plants. Grants of this sort continued into the 1980s because of commitments made before high oil prices obviated the need to subsidise. The remainder of the income pays for part of the difference between domestic and import prices. However, only enough to aid 11 million tonnes of consumption are available. Düngen (1984) reports that of the 12,000 million Deutschemarks raised by the tax from 1979 to 1984 about 9,000 went to subsidised coal output, 2,000 to powerplants investment, and the rest to other uses.

Compulsory long-term purchase contracts have been adopted to force further purchases. Contractual commitments to purchase German coal are made to increase electric utility coal use. The first such contract nominally to settle purchases for the next ten years was signed in 1977. It committed utilities to buy 33 million tonnes per year. A revised accord was reached in 1980 and obligated the industry to buy 640 million tonnes of hard coal from 1981 to 1995. The purchases were to be at a 38 million tonne level from 1981 to 1985; 43 million from 1986 to 1990 and 46 million from 1991 to 1995.

Tied to this were the rights to increase coal imports. From 1981 to 1987, German utilities could import a tonne of coal for every two tonnes in excess of 33 million that they employed. From 1988, an additional tonne of imports could be allowed for every tonne in excess of 33 million tonnes of German coal purchased. The deal was dubbed "Der Jahrhundertvertrag," literally the century contract but more aptly the contract of the century.

Given the uncertainties about demand growth and the availability of other energy sources, the compatability of the contract with other plans is unclear. Obviously, the lower the generation and the higher the nuclear output the less room remains for coal. Thus, if electricity output growth is low but nuclear generation expansion is high, it is unlikely that demand will suffice to absorb the contracted amounts of German coal, the allowed imports, and present levels of lignite use. Even with higher demands, it may be difficult to absorb these amounts of coal. It may be the contract rather than the other purchases that are altered. In 1985, this coal cost $25–$35 more per tonne than imported coal.

By all reckoning (except perhaps by German supporters of the industry), the Germans have the most persistent commitment to coal. Discussions in Germany evoke an image of a historical tradition to be preserved. The fervor matches that of Hans Sachs' (in Wagner's *Die Meistersinger*) plea for sacred German art.

Coal in Belgium[3]

As noted, Belgian coal was problem plagued well before the 1958 turnabout in energy expectations. Special programs were instituted to cushion the impact of joining ECSC. As small as it is, Belgium is split between a French speaking Walloon community in the South and a Dutch speaking Flemish community in the North. Tensions among the two long have been a major problem. Belgium has moved to creating institutions to give greater autonomy to the two communities.

The Belgian coal industry started in the South and consisted of many small mines. Early in the twentieth century, mines were developed in the Campine in the North. Total output of seven Campine mines peaked at 10.5 million in 1956.

At the start of ECSC, concern concentrated on protecting the southern Belgian mines. The resistance limited output declines in the early years of ECSC. Southern output only fell from 20.6 million in 1953 to 18.7 million in 1957. Then, a prolonged phase out was instituted in the 1960s. By 1965 output was down to 10 million; by 1970, 4.2 million; by 1975, 1.5 million. In 1984, 102,000 tonnes were produced by the last mine before it closed. This closure lagged three years behind the target set in 1975 for closing the South by 1981. In contrast, the 1975 goal of stabilising Campine output at

7 million tonnes proved unsustainable. Southern interests turned to protecting other industries.

The Campine mines began to suffer with the post-1958 coal crisis. Great effort was devoted to preserving them. In 1967, a single company was created to run the mines. By then only five were operating because of a merger of operations in 1964 and a closing in 1966. At the time of the merger, output was almost 9 million. Since 1973, output has been within a few hundred thousand tonnes of 6 million tonnes. Another mine consolidation was announced in 1985.

Moreover, the large and growing losses appear to be creating pressures to close Campine capacity. The rise of regional autonomy might even lessen the support of subsidies. Previously, the North took subsidies as the quid pro quo for support of aid to southern industries. When the North has a choice over use of a fixed budget, other priorities might dominate.

The great bulk of the assistance has been through explicit subsidy. Only limited amounts of forced purchases of coal are at above market prices. For most of the sixties and into the early seventies, a coal directorate operated to coordinate the intervention. Starting in 1980, the state decided to consider its subsidies to the Campine mines as acquisition of equity. The ownership share started around 70 percent and mounted to about 77 percent by 1984.

French protection

As noted, the French have moved towards substantial contraction of coal output with stress on the coking coal of Lorraine. Lorraine is near the traditional centers of steel making. Apparently lower costs, better coal quality, and location make preservation seem attractive. Substantial efforts are made to create new jobs. A critical influence is stress on developing nuclear power.

Two key institutional features have already been noted – the nationalisation of coal output and electricity. Another factor is that CDF operates powerplants. The price EDF pays for coal can be above world prices or EDF can buy power from CDF at costs that exceed the cost of generating with imported coal. However, knowledgeable observers outside the two organisations indicate no concessions actually are given.

Another key institution in French coal is Association Technique de l'Importation Charbonnière (ATIC). ATIC conducts and at least nominally controls the purchase of imported coal in France. Given that EDF and CDF have become the primary users of imported coal, it is unclear how much independence ATIC possesses. (9 of the 17 million tonnes of coal imported into France in 1984 went to EDF.) Control is greatest over coking coal used directly by steel mills. One of the major retreats of the Mitterand government was from its pledge to raise coal output to 30 million tonnes. The government decided instead to accept its predecessor's plan to lower output to 10 million tonnes by 1990.

The cost of protection

The prior discussion indicates that aid to coal involves far more than explicit subsidies. Many costs are hidden or at least not fully reported. Thus, data are not reported on the extra costs utilities, particularly in Germany, incur buying coal at unsubsidised prices above world levels. The EEC reports kohlepfennig receipts as a separate item since they are outside the formal subsidy system. Given the difficulties in determining an appropriate interest rate charge, the amount of interest subvention is difficult to estimate.

The outlays in any given year simultaneously make up failures to set aside money to cover current pension and pollution control expenses resulting from past production, cover losses on current output, and invest in future output. Moreover, many subventions, in practice, are given only to part of production. Thus, the EEC subsidy reporting system provides only rough indicators of the cost of aiding current production. The figures are incomplete since only direct outlays are treated.

The Community data are not designed to analyse how much of the cost of past mining could have been avoided by prompter rationalisation of the industry. The division of subsidies to production by total tonnage to give costs per tonne is at best suggestive. In the German case, the two key direct subsidy programs, coking coal sales and the kohlepfennig, each assist only a part of sales and thus the cost per tonne *aided* is much higher. Britain reports much lower pension obligations on the ground that no special programs are involved.

As Table 7.1 shows, the EEC distinguishes many subcategories of aid. The data are as reported by EEC in its European currency unit.[4] As the table shows, the value of the unit in terms of any one currency is variable. Over the period of study, the value has averaged about one US dollar. Significant increases have occurred in all the main classes of subsidies over the period studied here, namely all years in which data are in ECU. The costs of past production significantly exceed the aid to current output. By any standard, the Germans incur the highest total subsidy costs of the countries involved. The composition of German aid also differs radically from that elsewhere. Germany stresses sales promotion and has the lowest per tonne coal subsidies. However, kohlepfennig receipts roughly equal direct aid. The sales aids for Germany go predominantly into coking coal. A third form of German aid is the forced purchase under the contract of the

Table 7.1 *European subsidies of the coal industry and their main components 1976–85*

	1976	1977	1978	1979	1980	1981	1982	1983	1984	1985
a. Total payments to aid production in millions of ECU										
Germany	252	380	898	1,162	1,155	755	710	954	876	1,065
German electricity tax	NA	560	692	956	933	745	782	881	893	942
Total German	NA	940	1,591	2,118	2,088	1,500	1,537	1,591	1,769	2,007
Belgium	166	212	241	308	303	261	175	161	214	231
France	297	414	482	482	467	424	574	557	558	568
UK	27	55	185	276	294	856	723	1,492	3,783	NA
Total direct subsidy	742	1,060	1,806	2,228	2,218	2,295	2,182	3,164	5,432	NA
Total plus German tax	NA	1,620	2,498	3,184	3,151	3,040	2,964	4,046	6,325	NA
US dollar value	NA	1,848	3,183	4,364	4,388	3,394	2,904	3,601	4,992	NA
b. Subsidies of coal industry losses in millions of ECU										
Germany	14	0	0	0	0	0	0	0	0	0
France	291	407	474	474	458	415	565	548	543	5
UK	0	0	77	227	244	793	683	1,471	3,771	NA
Belgium	113	132	136	169	138	101	78	100	103	107
EEC Total	418	539	687	870	841	1,309	1,325	2,119	4,417	NA
c. Coking coal sales subsidies in millions of ECU										
Germany	34	172	496	741	752	301	304	624	590	583
France	1	1	2	1	2	2	2	1	1	0
UK	3	12	12	12	0	0	0	0	0	0
Belgium	30	64	80	107	144	137	74	34	84	98
EEC Total	68	250	590	861	897	439	380	659	675	681
d. Subsidies of investment in millions of ECU										
Germany	119	94	311	327	279	298	238	137	112	92
Belgium	0	0	7	13	11	11	12	13	13	13
EEC Total	119	94	318	340	290	308	250	150	125	105
e. Production aid in ECU per tonne of total output										
Germany	2.6	4.2	10.0	12.5	12.2	7.9	7.4	10.6	10.3	12.5
German electricity tax	NA	6.1	7.7	10.3	8.8	7.8	8.1	9.4	10.5	11.0
German with elect.	NA	10.3	17.6	22.7	21.1	15.7	15.5	20.1	20.8	23.5
Belgium	23.1	29.8	36.5	50.5	48.1	42.7	27.0	26.4	34.0	35.0
France	13.6	19.4	24.5	25.9	25.8	22.8	34.0	32.8	33.6	36.9
UK	0.2	0.5	1.5	2.3	2.3	6.8	6.0	12.8	76.4	NA
EEC Average	3.0	4.4	7.6	9.3	9.0	9.3	9.1	13.8	34.5	NA
f. Subsidies of social security outlays due to past mining in millions of ECU										
Germany	1,762	2,137	2,383	2,508	2,651	2,682	2,821	3,054	2,954	3,058
Belgium	459	550	601	629	636	704	745	925	943	959
France	849	942	1,080	1,202	1,341	1,418	1,504	1,515	1,483	1,468
UK	48	43	56	52	67	104	105	104	103	NA
EEC Total	3,117	3,673	4,120	4,390	4,696	4,908	5,175	5,597	5,482	NA
g. Subsidy of other inherited costs in millions of ECU										
Germany	150	198	203	203	187	188	133	108	123	70
Belgium	1	1	0	1	0	0	0	0	0	0
France	51	58	71	79	98	97	107	304	322	395
UK	16	19	22	27	28	59	202	421	446	NA
EEC Total	218	275	297	309	313	344	442	833	892	NA
h. Value of ECU										
US Dollar	1.12	1.14	1.27	1.37	1.39	1.12	0.98	0.89	0.79	NA
German Mark	2.82	2.65	2.56	2.51	2.52	2.51	2.38	2.27	2.24	NA
French Franc	5.34	5.61	5.74	5.83	5.87	6.04	6.43	6.77	6.87	NA
Belgian Franc	43.17	40.83	40.06	40.17	40.60	41.29	44.71	45.44	45.44	NA
British Pound	0.62	0.65	0.66	0.65	0.60	0.55	0.56	0.59	0.59	NA

Source: Subsidies: European Communities, 1985, *Statistical Data on Trends
in the Community Coal Industry Since 1975*, SEC(84) 1584.
European Currency unit: European Communities Statistical Office.
Basic Statistics of the Community, 1985, p. 66-67.

Note: Excess of German 1985 total over parts in original
1983,1984, and 1985 are projections to EEC.

century. At least 25 million tonnes per year of these purchases receive no aid from the kohlepfennig. Given the 40 ECU/tonne excess cost, at least one thousand million ECU of further aid is provided. Conversely, the British and French provide the bulk of their aid as a subsidy of coal industry losses. Finally, Britain is an exception to the rule that the debts inherited from old operations are more burdensome than aid to current output. Britain claims only limited *special* funding is needed to cover the costs of miners' pensions. Other types of inherited costs, however, have risen sharply in Britain.

As previously emphasised, these data are imperfect indicators of the costs of protection. However, they suffice to confirm the basic point that thousands of millions are being spent to protect coal mining jobs. Given German employment of 172,000 (Statistik der Kohlenwirtschaft), a 3,000 million ECU outlay is over 17,000 ECU per worker; Britain's 1,000 million ECU and 191,300 workers (NCB) is about 5,000 ECU per job. France's 550 million for 56,000 workers (CDF) is almost 10,000 ECU. The Belgian 200 million ECU for 18,000 workers (Fédération Charbonnière) is about 11,000 per job per year.

While worse cases of measurable subsidies exist, these figures probably greatly understate the costs. Many subsidies are left out. Moreover, the average undoubtedly conceals a considerable dispersion of costs. Some workers in the best mines may be generating profits that the firm uses further to subsidise output. The worst mines then will incur per tonne per worker costs far above average.

The best data on cost dispersion are those provided for Britain by the 1983 report of the Monopolies and Mergers Commission. The figures indicate that the worst mines have more than double the cost of the average mines. With average costs of 90 ECU per tonne, the worst mines would thus require over 100 ECU (the 90 ECU in above average costs plus the 18 ECU per tonne of average aid) in aid. This more than fivefold increase in aid to the marginal mine surely translates into an even greater cost per worker difference. Mines are costly because working conditions produce low output per man day. In any case, costs per worker in the worst mines are unlikely to be less than 25,000 ECU and probably are closer to 50,000.

Similarly, observers of the German industry suggest the worst mines have costs 25 to 50 ECU per tonne above the average. The average subsidy including the kohlepfennig and the cost of the contract of the century are about 30 ECU per tonne. Thus, the subsidy on high cost output may be 50 to 75 ECU per tonne. The cost per worker then is at least double the average and could easily be four or five times larger. Thus, 35,000 to 85,000 ECU per worker per year may be spent on preserving jobs in the worst mines.

Charbonnages de France now reports losses of individual mines. The worst ran losses that exceeded the CDF average by over 40 ECU/tonne. If this is added to the 35 ECU/tonne subsidy, subsidisation in the worst mines is at least double the average. Thus, subventions of 20,000–40,000 ECU per worker per year may occur in the worst mines.

The persistence of costs of such magnitudes leads to periodic decisions to reduce the amount of capacity supported. These are indicators of an unwise policy that should be more radically reformed.

Pollution control impacts

For numerous reasons, review here of air pollution issues in Western Europe is limited to noting that the concerns impose further limits to coal use.[5] Regulations of various degrees of stringency have been imposed. Both the European Communities and the UN Economic Commission for Europe have proposed regional programs of emission reduction. However, considerable difficulty has occurred in securing accord. Among coal producers, Germany has moved to the most stringent regulations. Britain has been resistant to the European initiatives, insisting among other things that the formulas calling for a fixed percent reduction in emissions inadequately consider past pollution control efforts.

Conclusions

Europe has a strong devotion to protecting the coal industry. The present discussion has only provided incomplete indicators of the magnitude of the aid. Since this is an introductory study of coal issues rather than of public policy in general, the question of how these aids compare to other subsidies in coal, energy, or elsewhere was not addressed. Comparisons were not even made to other programs to aid coal such as in Japan and eastern Canada. The goal was to show the programs were undesirable, not to compare their undesirability to the many other ill-conceived policies prevailing. Almost certainly the money invested could have been better used to encourage development of profitable modern industries to employ the workers kept in the coal industry.

From the rise of the coal crisis, observers have warned of the perils of subvention. They were mocked for not recognising that higher oil prices would come and make coal use necessary. In fact, even oil price rises did not help. Nevertheless, support of coal policy remains strong.

CHAPTER 8

The emergence of producers for export

Particularly in the 1970s, several countries have emerged as major participants in international trade in coal. Australia became the world's leading coal exporter. South Africa developed into another major participant. Canada also has an important role. In 1985, Colombia entered the market. Conjectures and even tentative development plans have arisen about other entrants including mainland China, Indonesia, and Botswana.

This chapter stresses the development of the Australian and South African coal industries. Note is taken of Canada and the Colombian venture. Stress again is on the complex business–government relations at work.

Among the producers for export, their differences are at least as considerable as their similarities, but several similarities exist. The most obvious are their newness as suppliers, the critical role in supply of large diversified companies, the importance of large open pit mines, and the tendency to construct new specialised port facilities. Such ports can receive coal from unit trains, move it from trains to storage areas to ships with highly mechanised systems, and accommodate large ships.

Australia to date has exported far more than the others. These exports, in fact, reached a level in 1984 that allowed Australia to surpass the United States as the world's leading coal exporter (recall Table 3.3). However, Australia had not yet matched the US levels of the latter's best years. Conversely, Australian exports significantly exceed that nation's total "black" coal consumption (see Table 8.1). The trade started with exports of coking coal to Japan, and this remains the largest component. The rise of steel production elsewhere, notably Korea and Taiwan, has created new markets. So has a rise in steam coal trade (see Table 8.2).

While Australia seems to favor substantial coal exports, the South African government has nominally tried to regulate export levels. Although the underlying economics are complex, the reticence, in practice, seems form without substance. The South African government nominally sets

Table 8.1 *Selected data on Australian coal, 1950–85 (million tonnes or percents)*

	1950-51	1960-61	1970-71	1984-85
Total salable output NSW	12.9	17.4	31.1	58.3
Total salable output Queensland	2.3	2.6	11.1	54.3
Total salable output Australia	16.7	22.2	45.1	118.3
NSW percent of Australia	77.2	78.5	68.9	49.3
Queenslands percent of Australia	13.7	11.9	24.7	45.9
Two states' percent of Australia	90.9	90.4	93.6	95.2
NSW deep raw coal output	13.7	17.6	33.1	42.1
NSW opencast raw coal output	1.1	0.8	2.6	28.0
Percent opencast NSW	7.4	4.4	7.2	39.9
Queensland deep raw coal output	2.1	2.6	4.5	5.3
Queensland opencast raw coal output	0.7	0.4	9.9	63.9
Percent opencast Queensland	24.2	13.8	68.8	92.4
Australian deep raw coal output	16.9	21.2	38.2	48.6
Australian opencast raw coal output	2.7	2.4	14.8	96.5
Percent opencast Australia	13.8	10.2	28.0	66.5
NSW percent of Australian deep output	81.3	83.0	86.8	86.5
Queensland percent of Australian deep output	12.7	12.4	11.7	10.8
Two states' percent deep output	93.9	95.4	98.5	97.4
NSW percent Australian opencast output	41.4	33.9	17.5	29.0
Queensland percent Australian opencast output	25.3	17.4	66.5	66.2
Two states' percent opencast output	66.7	51.3	84.0	95.2
Exports from NSW	0.1	1.9	12.0	38.3
Exports from Queensland	0.0	0.0	7.0	45.5
Total Australian exports	0.1	1.9	19.0	83.8
Australian coking coal exports	0.1	1.8	16.9	50.5
Australian steam coal exports	0.0	0.1	2.1	33.3
NSW percent of exports	100.0	97.5	63.2	45.7
Queensland percent of exports	0.0	2.5	36.8	54.3
Coking coal percent of exports	80.9	95.6	89.0	60.3
Steam coal percent of exports	19.1	4.4	11.0	39.7
Total Australian coal consumption	17.1	20.0	25.0	40.7
NSW Consumption percent	61.9	68.3	68.8	56.9
Queensland consumption percent	13.3	13.0	14.7	25.7
Electricity percent of Australian coal use	26.4	35.0	50.9	73.1
Consumption and exports NSW	10.7	15.6	29.2	61.4
Export percent in NSW	0.6	12.1	41.1	62.3
Consumption and exports Queensland	2.3	2.6	10.6	56.0
Export percent in Queensland	0.0	1.8	65.6	81.3
Consumption and exports Australia	17.2	22.0	43.9	124.5
Export percent in Australia	0.4	8.8	43.2	67.3

Source: Joint Coal Board, *Black Coal in Australia* 1982-83 and 1984-85 except 1950-51
by mining method from direct communication from Joint Coal Board.

coal export quotas on the basis of what can be allowed without creating threats to secure domestic energy supplies. However, this ceiling has about the same substance as the floors European governments set on production. The South African government has increased the export limit three times. In every case, industry requests based on estimates of what could be marketed were accepted. Most South African coal is steam quality, but a few million tonnes of specially prepared and blended cokable coal are sold to Japan.

Table 8.2 *Australian trade 1978–79 versus 1984–85 (thousand tonnes)*

	Australia 1978-79	New South Wales 1978-79	Queensland 1978-79	Australia 1984-85	New South Wales 1984-85	Queensland 1984-85
Total	38,278	19,442	18,836	83,799	38,296	45,503
Coking	33,257	14,623	18,634	50,523	16,142	34,381
Steam	5,021	4,819	202	33,276	22,154	11,122
Japan total	25,174	12,265	12,909	43,923	19,328	24,595
Japan coking	24,354	11,483	12,871	31,375	11,714	19,661
Japan steam	820	782	38	12,548	7,614	4,934
Europe total	8,834	4,052	4,782	19,326	6,739	12,587
Europe coking	5,209	485	4,724	10,201	751	9,450
Europe steam	3,625	3,567	58	9,125	5,988	3,137
S. Korea total	1,933	1,464	469	7,215	5,310	1,905
S. Korea coking	1,933	1,464	469	3,050	1,687	1,363
S. Korea steam	0	0	0	4,165	3,623	542
Taiwan total	1,362	1,055	307	4,509	2,736	1,773
Taiwan coking	1,126	899	227	1,661	671	990
Taiwan steam	236	156	80	2,848	2,065	783

Source: Joint Coal Board, *Black Coal in Australia*, 1982-83 and 1984-85.

Note: Europe here includes Rumania which received 145,000 tonnes of NSW coking coal
in 1978-79 and 1.04 million tonnes in 1984-85.

The cheapest to produce Canadian coal is located in the western provinces. The principal Canadian coal users are in Ontario and nearer to US than to Canadian mines. Thus, the US supplies Ontario. The western province mines have developed an extensive export trade and some local sales. Colombia started significant exports in 1985 as a large new surface mine began commercial operations.

Appraisal of the relative competitive position of different countries involves considerable difficulties. Much interesting work has been done on estimating coal supply curves for different countries. In some cases, these estimates can be based on informed analyses of the costs of the small number of large mines operating in the country. In other cases, such as Appalachia in the United States, too many diverse small mines operate to permit analysis of all their costs. More critically, in the latter case, long term supply depends to a great extent on expansions that are more difficult to identify than in areas in which large projects that take several years to complete produce the expansion. Available analyses invariably are done by those outside the producing companies and so involve use of imperfect data. At best then we get reasonably accurate estimates of current supply curves. Less confidence can be placed in the ability to project how depletion, new mine openings, technical progress, and wage and other costs will alter supply.

A final difficulty in making international comparisons is predicting exchange rates. The dollar has fallen, risen substantially, and fallen again. The South African rand is vulnerable to political instability in that country. Exchange-rate-change effects are complicated by their impacts on mining equipment costs. US companies are the prime source of surface mining equipment with Japanese competition for earth moving equipment, large trucks, and other machinery designed for heavy construction as well as mining. Underground technology is sold by European and US companies. International financial theory stresses that part of exchange rate variation merely reflects differential inflation and leaves competition positions unchanged. However, the swings since the abandonment of fixed exchanges have been considerably greater than can be explained by price-level changes. Real changes in competitiveness have occurred. The possibility of further changes must be considered. In particular, any comparison of countries must consider the possibility of exchange rate movements greatly altering competitive position.

Coal in Australia

Australian sources distinguish sharply between black coal (i.e., hard) and lignite. The former is predominantly produced in New South Wales (NSW) and Queensland. Lignite is produced in Victoria by the state electricity board. Thus, the statistical separation reflects a real distinction in both location and the identity of the participants.

The two main Australian black coal producing provinces differ greatly in both their general economic development and the nature of their coal industries. NSW is the most populous industrialised state with Sydney as its capital and principal city. Queensland is much less developed. Victoria, whose capital and population center is Melbourne, is another highly developed state.

The coal industries in two northern states, which long have together produced over 90 percent of Australian black coal output, differ in several critical ways.[1] These include differences in reliance on exports, coking coal, and surface mining. Coal production developed in different ways in each state. Key indicators of these trends appear in Table 8.1.

Australia became a significant coal producer only after World War II. New South Wales developed sooner and remains the larger producer, but its growth has lagged behind Queensland. Queensland is more export, surface mining, and coking coal oriented than New South Wales. However, this change in position did not start until the 1960s. Through the 1950s,

NSW production grew at about the same rate as national production. In the middle sixties, substantial export-led growth began in Queensland and has continued. After 1974–75, growth in the two states became more similar.

A second major difference between the two relates to mining methods. Both have moved to increased reliance on opencast (surface) mining. However, Queensland moved to opencast sooner than NSW and produces much more opencast coal in terms of amounts, share in Australian output, or share in regional output.

The final critical distinctions relate to disposition of coal. Since 1973–74, total Australian exports have exceeded internal consumption. Export dependence is far greater in Queensland than in New South Wales. Queensland began exporting more than it consumed in 1968–69. In contrast, 1982–83 was the first year in which NSW exported more than it used domestically.

The Australian coal industry's greater dependence on exports than on internal sales is unique among coal producers. Exports have been rising for many decades (see Tables 4.3, 8.1, and 8.2). The position of the two main producing states has altered considerably. New South Wales was exporting more than Queensland through 1971. From 1972 to 1981, Queensland exported more than New South Wales. The latter overtook Queensland in the next two years but lagged behind in 1984.

While coking coal remains more important than steam coal and Japan remains the largest customer, steam coal sales have greatly increased in importance in the 1980s, new customers have emerged – notably Western Europe, South Korea, and Taiwan.

In many cases, Australian coal mine ownership is shared with one partner dominant. Including the shares of their partners, the eight leading firms control about 80 percent (see Table 8.3). Broken Hill Proprietary (BHP) has become by far the largest producer. It is the largest owner of two major operations in Queensland – those of Utah International and Theiss-Dampier-Mitsui (TDM). In both cases, the operations were acquired from US companies. The smaller TDM purchase, from Peabody Coal, came in 1977.

This purchase was an offshoot of a US antitrust case that forced Kennecott Copper to sell its ownership in Peabody. In 1985, CSR, one of the partners, sold its share in TDM to BHP. Utah was one of the first, the largest, and the lowest cost of the Queensland opencast producers. In 1977, the US General Electric Company bought Utah. While this apparently was one of the more successful mineral company acquisitions, GE nevertheless decided in 1982 to sell Utah. GE felt Utah did not fit into GE's long term plans. By 1984, a deal was set up that transferred Utah to BHP and restructured Utah's Australian operations. Among the key elements of the

Table 8.3 *Production by companies in Australia fiscal year 1984–85*

	Raw Coal Output in Million tonnes	Percent National Total	Percent State Total
BHP Queensland	35.3	24.3	51.0
BHP NSW Wholely Owned	9.5	6.6	13.6
Blue Circle Portland Cement	0.7	0.5	0.9
BHP Total	45.5	31.4	
Elcom New South Wales	9.4	6.5	13.4
Elcom Contractors	5.5	3.8	7.8
Elcom Total	14.9	10.2	21.2
Howard Smith Queensland	3.4	2.3	4.9
Howard Smith NSW	10.3	7.1	14.6
Howard Smith Total	13.7	9.4	
CSR Queensland	5.3	3.6	7.6
CSR NSW Wholely Owned	3.0	2.1	4.3
CSR NSW Joint Ventures	1.7	1.2	2.5
CSR Total	10.0	6.9	
MIM Queensland	10.2	7.0	14.8
CRA Queensland	4.8	3.3	7.0
CRA New South Wales	2.6	1.8	3.7
CRA Total	7.4	5.1	
BP New South Wales	6.1	4.2	8.8
Shell Queensland (German Creek)	3.7	2.5	5.3
Shell (Austen&Butha) NSW	3.4	2.4	4.9
Total Shell	7.1	4.9	
Total Above	115.0	79.2	
Shell Joint with CSR Queenland	3.0	2.1	4.3
Shell Joint with CSR NSW(Drayton)	1.6	1.1	2.3
Shell Joints	4.6	3.2	
Shell Total	11.7	8.1	
Howard Smith for Elcom	1.6	1.1	2.3

Sources: Joint Coal Board,*Black Coal in Australia 1984-85*, pp. 34-38
  Queensland Coal Board, *34th Annual Report*,pp. 16-17 for output by mines,pp. 49-71 for ownership
Notes: In many cases, joint ventures are involved and the total production is credited here to the largest owner.
  While Queensland reports output by mine on a raw and cleaned coal basis, New South Wales presents only raw coal figures so these were used for both states.
  Shell owns 30 percent of Boundary Hill and Callide in Queensland;CSR 55 percent
  The Drayton shares are 44 percent CSR and 39 percent Shell.

restructuring were a temporary GE holding of position (liquidated in 1985) and integrating a separate BHP mine into the venture. (Prior to the takeover, Utah had taken Mitsubishi and an Australian insurance company as partners.) BHP also long has maintained operations in NSW that slightly exceed TDM in total output.

Australian sources agree Utah has low mining costs and substantial potential to expand output. Production, however, has been stable for several years. In contrast, BHP's NSW operations are higher cost. On a pure cost basis, it might be desirable to reduce BHP output in NSW and replace it with expansion of Utah. However, political resistance to mine closing would be a severe barrier to the move.

The other major participants in Australian coal production differ markedly in the nature of both their overall operations and their involvement in coal. The second largest producer is the Electricity Commission of New South Wales (Elcom). All its production is in NSW and is mostly for power plants. Two diversified firms, Howard Smith and CSR, are the next most important producers, and are involved in both NSW and Queensland, albeit in quite different proportions. Two other minerals companies, MIM and CRA, come next. MIM is only in Queensland; new mines have made CRA's output greater in Queensland than in NSW. Shell and British Petroleum are the other remaining major participants. BP's operations in NSW are with one small exception wholly owned. In contrast, Shell's holdings are all joint ventures including several in which it is a junior partner. As the data also show, while BHP produces the majority of Queensland output, production shares of leading companies are smaller and more similar to each other in NSW.

The two main features of government involvement in Australian coal are first in devising methods to transfer economic rents to government, and second, to mediate tumultuous labour–management relations. Railroad rates in Queensland and New South Wales apparently are the primary method of rent transfer. The situations in the two states are quite different. New rail lines were built for the new opencast, export mines of Queensland. The mining companies put up the capital. Thus, the cost–price relationships are much less muddled by the effects of other traffic. Australian observers argue the data conclusively show that as a way to tax the mines, the rates are set far in excess of costs.

In contrast, NSW coal traffic in large measure travel over tracks shared with a considerable amount of other traffic. The system suffers from congestion. Thus, it is difficult to determine exactly how much of the high rail rates endured results from high costs of operation and how much is an implicit tax.

Conventional taxes and royalties are also imposed. The federal government has introduced and modified taxes on coking coal exports. The stress has been on taxing low cost opencast producers of high quality coking coal, particularly Utah. Such coal was subject to a tax that at its height was $6.00 (Australian) per tonne. Since 1982, the tax has been at $3.50 and is only levied on the Queensland BHP mines; all other export taxes have been removed. Queensland imposes royalties of 5 percent of the free on rail value of surface mined export coal, 4 percent on underground mined export coal, and five cents per tonne on domestically consumed coal. NSW has a basic royalty of $1.70 per tonne and introduced extra royalties of $1.05 to $1.75 on new mines.

The Australian mining industry regularly contends the tax burden is limiting Australian ability to compete. Officials of at least the mining ministries at the federal and state level are sympathetic to these complaints. Efforts are being made to reappraise tax policies, stabilise rail rates, and improve labour relations.

Another critical concern is that labour relations are considered complex and tumultuous. Many unions are involved in the mining (and transporting) of coal. Dealing with so many unions by itself causes severe management problems. Militance of the unions aggravates the difficulties.

However, concern has shifted towards working out a program for rationalising the industry. Industry sources feel that too much capacity exists and that, in particular, it would be desirable to shift from underground to more opencast output in New South Wales. Such a move requires working out agreements with the affected unions.

Australians believe that they have the largest block of identified low cost coal in the world. Thus, a concept of a natural position of supremacy is enunciated. Government restrictions and restrictive labour practices then are deemed to undermine this natural strength. These statements reflect the overenthusiasm typical of industry advocates and neglect problems of currency fluctuation. However, outside observers generally share this optimism about Australia.

South Africa[2]

The critical features of South African coal include a long extant industry, extensive government involvement, and the pivotal role in coal mining of the large corporations that dominate South African industry. As part of its policy of lessening its vulnerability to external pressures, South Africa has encouraged continued heavy dependence on coal and promoted ventures to produce synthetic fuels from coal. Substantial coal production long existed but extensive efforts to export did not begin until the late 1970s. While only rough indicators are available on the role of coal, their review helps suggest the special situation of South Africa.

In the Department of Mineral and Energy Affairs report for 1981, oil use was reported to equal about a fifth of coal consumption. Coal consumption was 109 million physical tonnes so roughly 19 million TCE of oil was consumed. By 1984, coal use had risen (according to Chamber of Mines data that differ slightly from those from the Department) to almost 122 million tonnes. Oil use is likely to have at most risen in parallel with coal – i.e., to no more than 24 million TCE. The UN sets 1981 coal use for the

Table 8.4 *Patterns of coal sales in South Africa*

	All sales	Export	Total domestic	Electric	Escom	Total industrial	Synfuels	Industry net	Coking	Other	Trans-portation	Mines
1963	41.7	0.9	40.8	19.8	14.7	6.7	NA	NA	3.7	3.0	6.2	1.6
1965	47.8	0.8	47.0	22.8	16.7	7.9	NA	NA	4.8	3.7	6.6	1.3
1970	54.5	1.3	53.2	29.5	21.6	8.8	NA	NA	5.0	3.7	5.1	1.2
1975	69.0	2.3	66.7	41.3	34.2	12.1	NA	NA	5.2	3.8	3.6	0.7
1980	117.6	28.4	89.1	58.9	46.8	16.1	6.2	9.9	7.5	3.8	2.0	0.9
1983	143.1	30.1	112.9	60.1	55.0	39.6	23.9	15.7	7.3	3.9	1.2	0.8
1984	159.6	38.1	121.5	62.6	58.7	43.1	30.6	12.5	8.1	5.7	1.1	0.9

Sources: Except for Escom and Sasol,1963-80 South African Fuel Research Institute
1981-84 Department of Minerals and Energy Affairs.
As reported in the Chamber of Mines, *Statistical Tables,* Johannesburg
Escom and Sasol from annual reports; Sasol average of fiscal year output figures.

South African Customs Union at 81 million TCE; oil, at 15 million TCE; the respective 1983 figures are 83 million TCE and 16 million TCE. Thus, while details differ, the critical point is clear; South Africa is far more coal dependent than OECD countries.

The pattern noted above for other countries of dominance of electric power in coal use also prevails in South Africa. However, South Africa has a unique venture into synthesis of fuels and chemicals from coal. Rapid expansion of that sector in the 1980s has made it the second largest coal consuming category. Other uses are scattered. Industry, coking, and other uses have shown growth (but to a much less relative and absolute degree than electric power). Rail and mine use has declined (see Table 8.4).

Coal production is dominated by what are somewhat anachronistically called the mining houses – large diversified firms that began as mining ventures and have expanded into many other areas. They are connected through ownership of each other's stock and by marketing through joint sales agencies.

The government seeks to control monopoly by acting as a major buyer, by having some government corporations operate their own mines, and by regulating all domestic sales to the private sector. The largest customer is the state-owned Electricity Supply Commission (Escom). It relies on purchases from private industry, largely from mines dedicated to a power plant (see below). The Sasol Corporation for synthetic fuels – largely owned by a government corporation – owns and operates its own mines.

The state-owned South African Iron and Steel Industrial Corporation (ISCOR) has both captive and open market supplies but has been increasing reliance on captive mines.

The mining houses

The mining houses dominate mining and are far more closely linked than would be possible under antitrust laws in the United States. About seven legally distinct mining houses are significant entities in the coal industry. However, two companies – Anglo-American and General Mining Union Corporation (Gencor) – are much larger companies and coal producers. Anglo-American, moreover, is the most important house (see Table 8.5).

Anglo-American is intricately structured and is a significant stockholder in at least four of the other leading mining houses involved in coal mining. In one case, Johannesburg Consolidated Investment (JCI or Johnnies), Anglo lists its interest as controlling. However, JCI's management appears to possess considerable independence. The other holdings are in Gencor, Barlow-Rand, and Goldfields.

An emerging feature of the South African coal scene is the participation of international oil companies. The first involved were Shell, British

Table 8.5 *Coal sales and electricity generation in South Africa in 1984*

(million tonnes except as noted)

Parent	Subsidiary or source	Million Tonnes	Percent of Total	1984 Escom coal burn company mines	Coal sales to new Escom plants
Anglo-American		37.3	23.5	21.3	40.0
Gencor	TransNatal	36.2	22.8	22.7	
Barlow-Rand	Per Chamber	9.8	6.2	23.0	8.0
Duhva	Per Minerals Bureau	8.7	5.5	9.5	
Barlow-Rand	Adjusted-wholly owned	19.3	12.1		
JCI	Tavistock	4.6	2.9		
Lonrho	Duiker	3.1	1.9		
Kangra		2.8	1.8		
Goldfields	Apex	2.3	1.5		
Gencor/Goldfields	Clydesdale	4.2	2.6	2.7	
Iscor		2.0	1.3		11.0
Mines Reported by Chamber		102.5	64.4		
Duvha	Per Minerals Bureau	8.7	5.5	9.5	
Sasol	Per Minerals Bureau	27.9	17.5		
Rietspruit*	Per Minerals Bureau	4.8	3.0		
Middelburg#	Per Minerals Bureau	3.0	1.9		
Total above		44.4	27.9		
Omissions		16.5	10.4		
Total Sales	Grand Total	159.1	100.0		

Sources : Sales by Companies: Chamber of Mines,*Statistical Tables* and
South Africa, Department of Mineral and Energy Affairs, *Operating and Developing Coal Mines in the Republicof South Africa*, Johannesburg,1985.
Escom data from Escom annual reports.
Coal source of plants from Escom and coal company annual reports.

Notes: Clydesdale, also half owner with Gencor of Matla producing 10.1 million for Escom, is shown separately; control passed from Gencor to Gold Fields on July 1,1984 Amcoal sales to planned Escom plants includes 15 million to plant reported by Amcoal but not in Escom's annual reports to 1984.
* A joint venture of Barlow Rand and Royal Dutch/Shell
#Owned 89 percent by British Petroleum, 5 percent Barlow Rand; 6 percent third company.

Petroleum, and Total (Compagnie Française des Petroles). In each case, the mode of entry was a joint venture with a mining house. Gencor established the Ermelo joint venture equally owned by itself, British Petroleum, and Total. Barlow Rand operates and jointly owns with Shell the Rietspruit mine and Middelburg mine with British Petroleum. BP owns 89 percent; Rand, 5 percent; a third company, 6 percent. JCI has a joint venture with Total. AGIP of Italy's ENI (Ente Nazionale Idrocarburi) is planning an independently developed mine.

The remaining domestic sales, at least for the mining house affiliates, are handled by two sales organisations – the Transvaal Coal Owners Association (TCOA) and the Natal Associated Collieries (NAC). An Anthracite Producers Association has ceased marketing for its members.

Viewing the limited market prospects and the world energy market in the late 1960s, the industry decided to explore development of a more significant export trade. Unregulated sales in the export market were expected to prove more profitable than domestic sales. Tied to the process was a proposal to build a highly mechanised coal handling facility on Richards Bay on the Indian Ocean. The facilities are designed to unload trains quickly, move the coals to the separate piles each exporter wishes to maintain, and then rapidly move and load the coal on ships.

The initial proposal and its extensions have required government approval. The principal barrier to allowing exports has been concern that South Africa was endangering its long-run energy security by lessening its ultimate ability to supply itself. The usual justifications for exports were made more enticing by various devices for "sharing" the benefits of exports with South African coal consumers. For example, mines serving Escom could sell their most desirable output for export and charge Escom less. (This should be considered more as saving Escom from government-imposed inefficiency rather than a subsidy. Prior regulations prevented the more rational policy of isolating and exporting the more valuable products and only using the lower grade coal that an electric powerplant can usually readily burn).

Over the 1970–82 period, the government successively approved four allocations of export quotas tied to development and expansion of the Richards Bay terminal. The first phase was in 1971; the second and third in 1973 and 1974; the fourth in 1982.

Two phases of the Richards Bay project had been completed by the early 1980s, and the first part of phase III was coming on line. Negotiations were being conducted for the first part of Phase IV. Phase I was finished in 1976; Phase II, in 1979; Phase III is expected to be done by 1987.

Phase I called for 12 million tonnes; Phase II doubled the allocations; Phase III added another 20 million tonnes to bring the total to 44; about 36 million more were allocated in Phase IVA bringing the total for Richards Bay to 70 million. A second part of Phase IV would bring the total to 80 million. Provision was made for export of another 7 million under special programs.

Each phase of allocations was somewhat different. Phase I was allocated directly to the three sales associations – with 10 million tonnes to TCOA and a million tonnes each to the other two marketing associations. Minor adjustments in the quotas were made in subsequent phases.

Most of the remainder of Phase II went to Shell (3.75 million tonnes), BP (1.25 million tonnes), and Total (1.25 million tonnes) – purportedly for being reliable suppliers of oil. Anglo-American's Amcoal got 2.5 million tonnes; Gencor, 1.25 million tonnes.

Phase III also was allocated mainly among the oil companies and the three leading mining houses. The effect was to give Amcoal and Gencor cumulative total allocations of 6 million tonnes each; Shell and BP, 5.5 million each; Total and Rand, 2.5 million each; Kwa Ngoma, a Gencor affiliate that is the owner of Zululand reserves, got the remaining 1.5 million.

The Phase IVA allocations were more fragmented and complicated. The allocations were in three parts. The bulk consisted of about 25 million of unconditional grants. Two smaller (about 4 million each) allotments were made. One was to divert shipments going either from the congested South African port at Durban or through Maputo in politically unstable Mozambique. The other tranche covered low grade coals not marketable domestically.

The majors again got the bulk of the allocations. Prior quota holding companies received a further 12.5 million tonnes of the main allocation (4.5 million tonnes to Gencor, 4 million tonnes to Amcoal, 2.5 to Rand, and 0.5 million tonnes to each of the three oil companies). Other large companies not previously holding quotas shared in over 6 million tonnes in new allocation (2 million tonnes to AGIP, 1.5 million tonnes to JCI, one million tonnes each to Anglovaal and Goldfields, 500,000 tonnes to Lonrho, and 350,000 tonnes to Kangra). Another 24 companies shared in about 5 million tonnes of new quotas for high grade coal exports, almost 4 million tonnes of rights to export low grade coal, and rights to shift exports of about 4 million tonnes to Richards Bay. Both the total allocations and their composition differed considerably. In fact, 15 different allocation patterns were devised.[3]

The representatives of the large companies argued strongly that the small producers could not efficiently utilise their allocations. These producers were seen as incapable of profitably delivering coal to Richards Bay or maintaining facilities at the terminal. Concerns about the latter problem were echoed by the staff at Richards Bay. The segregation of coal piles was considered a difficult task even when dealing with the present large shippers. The problems were expected to worsen greatly if very small shippers had to be accommodated.

While technologically quite correct, these arguments are economically irrelevant. Anyone familiar with quota allocations recognises that the critical question is what rights the quotas really convey. The actual value of the quotas depends critically upon how much transferability is allowed and the value of the quota rights. If value exists, it can be reaped by resale.

Obviously, the big companies would prefer a policy of nontransferable quotas so that the inability of small firms to export profitably would force surrender of the quotas and the costless government reallocation to the larger companies. The small companies will want to be allowed flexibility to make deals to insure that they can benefit even if they do not directly export. This conflict is predictable. It follows the standard textbook predictions (and their frequent empirical verifications) about the problems arising when the government gives away, on the basis of administrative discretion, valuable property rights.

What makes this debate unusual is that the quotas supposedly exceed the amount that can be profitably exported and thus should be valueless. The concerns may arise because the actual potential is greater, because of a desire to promote policy changes if quotas become scarce, or simply because of lack of reflection. This question and its obverse, why South African restricts exports, are discussed more fully below.

Coal production patterns

The imperfect available data on companies suggests that four entities constitute about 85 percent of the industry (see Table 8.5). The two biggest by far are Gencor and Anglo-American. Sasol's self-production for its synthetic fuels plants comes next. When the mines it operates for oil companies are considered, Barlow-Rand produces about as much as Sasol.

The data also disclose that Gencor, Anglo-American, and Rand are even more dominant as suppliers of Escom (Table 8.5). Prospects are for Rand and Anglo-American gaining while Gencor sales are static. Escom's 1983 annual report indicated that five new plants each requiring at least 10 million tonnes of coal were planned from 1985 on to the 1990s. The first two of these would be supplied by Amcoal; the next by a mine of ISCOR, the government owned steel company, and the last two by Rand. Amcoal reports it will serve a sixth plant still in the early planning stages.

Anglo-American and Rand (counting mines operated for oil companies) are also the major exporters – accounting for at least 10 million tonnes each. Gencor is probably nearer to a 5 million tonne level. The rest of exports comes mainly from other major mining companies such as JCI, Gold Fields, Lonrho, and Kangra.

About 30 million tonnes are involved in nonEscom electricity and the five categories of other uses. Industry sources in South Africa indicated that TCOA marketed about 18 million tonnes in 1982; Amcoal's 1986 annual report indicates 1985 TCOA sales of 16.2 million tonnes. Thus, TCOA has a large part of (the rather stagnant) open market.

Coal export patterns and economics

South Africa started its export trade with a coking coal deal with Japan, but limits on economic supplies of coking coal have caused the bulk of the expansion to come from steam coal exports. These, moreover, have gone predominantly to Western Europe (see Table 4.3).

When I visited South Africa in 1983, industry sources argued that a recovery of coal prices to about $40 per tonne or 45 Rand at the then prevailing exchange rate was needed to justify the export expansion program. Subsequently, the sharp depreciation of the Rand has severely altered the situation. Required Rand prices presumably have risen substantially but probably by less than the Rand has depreciated. Thus, a price recovery is no longer critical. In any case, the industry sources may have overstated their price rise needs. Thus, the best guess is that South Africa can produce the Phase IV level of exports at costs that are competitive with those in Australia. The critical problem is political instability.

Considerable unanimity of opinion existed among South African coal industry sources about the prospects for coal exports and for additional quotas. The expectations of a booming coal market propelled by continually rising oil prices no longer prevailed. The actual prospects are much less.

Internationally, the South African producers saw modest prospects to serve the electric power market. Specifically, the general expectation is that the plans to expand export capacity are sufficient to serve South African export sales growth into the middle 1990s. This view is based upon expectation of securing a healthy share of a modestly expanding market. Growth of this sort means that for the foreseeable future exports will be below both the allocations already made and thus to the 100 million tonne level that industry sources fear would trigger domestic objections to further rises. They also recognise the obvious conclusion that this means that the 100 million tonne benchmark has no practical relevance. At best, it is dubious to speculate how attitudes will evolve over the long period before the industry actually pushes exports to what the industry considered the present upper limit to quota levels.

Another point on which accord is widespread is that Escom is likely to be the only domestic growth market. As suggested, much of the growth into the early 1990s (and possibly beyond if slippages in expansion continue) has already been assigned to existing firms – mainly Amcoal and Rand. Beyond this, concern exists that only Gencor and the oil companies are also large enough to operate mines of the size Escom requires. (The mines each produce much more than the total output of any other producer. However, experience in the United States suggests that a well managed company with access to the capital markets can undertake Escom-sized ventures.)

Thus, South Africa views its coal prospects with cautious optimism.

Conclusions on South African coal

This preliminary view of the South African coal industry has disclosed the tendency to regulation that plagues energy policy everywhere. As usual, the situation is rife with contradictions. A country that invests extensively in synthesising petroleum from coal also tries to overstimulate direct use of coal with price controls. The efficacy of these controls is unclear. Examination of coal consumption patterns suggests that at best 15 percent of the domestic market might require subsidy to continue coal use (10 million tonnes in industry and 4 million tonnes in retail sales).

The export quota allocation process is another example of an all too familiar affect of regulation – an arbitrary allocation of rights.

The vigor of competition in South African coal is far more difficult to appraise than in any other major producing country. Indicators of monopoly potential, notably the dominant role of large firms, high profits, and the fight for quotas, suggest monopoly power. Yet, the South Africans are viewed by producers elsewhere as vigorous competitors on the world market, and South African producers see themselves as limited most by market forces. The temporary bottlenecks at Richards Bay cause limitations on competition that can be lessened by expansion.

Several explanations are possible. One suggested above is that the controls are ineffective and the firms overestimate the values of holding quotas. The comments about market limits would then relate to what could be profitably sold in a competitive market.

Alternatively, South Africa might indeed possess and utilise monopoly power in the world coal trade. The increases in the quota could be interpreted as the response of the monopolist to changing demand. Quotas then become valuable for the usual reasons.

An intermediate view is that physical limits on the pace at which one can efficiently expand Richards Bay and the rail lines to it from mines govern the behavior. Rents are temporarily available during these development phases. These temporary rents then are what makes quotas valuable. This last view also better explains why eventually significant price cutting occurs. However, this last point is not conclusive. Even a monopolistic South Africa might find price cuts desirable, given the large number of rivals that could be displaced. The biggest objection to a permanent monopoly view is that so much low cost coal seems available in the world.

Canada[4]

Canada ranks fifth in tonnage behind Australia, the USA, Poland, and South Africa among coal exporters. Table 4.3 shows the heavy dependence on Japanese sales and the growing role of exports to Europe. A factor excluded from the table is that Korea and Taiwan also have emerged as significant customers.

Again, the industry structure is complex and full of national peculiarities. The leading producer is Westar (about 7 or 8 million tonnes); now controlled by the British Colombia government, it was started by Kaiser Resources. Another leading company (4 million tonnes) is Fording, jointly owned by the Canadian Pacific Railroad and Cominco, a mining company whose majority stockholder is Canadian Pacific. Luscar, another Canadian mining company accounts for about 9 million tonnes; 2 million of which is in a 50/50 venture with Consolidation Coal of the USA. Other participants include Exxon and Shell.

Shell is one of several companies opening new ventures in the early 1980s. Its Line Creek Mine started in 1983 with capacity of 3 million and produced 2.3 million tonnes in 1984. The largest new venture, with a capacity of 6.3 million tonnes, was Quintette also opened in 1983. Its half owner is Denison Mines, another Canadian company. Charbonnages de France owns 12 percent; various Japanese interests, 38 percent. Two other mines – Gregg River and Bullmoose – roughly similar in size to Line Creek also opened in 1983. In both cases, Canadian companies hold the majority share and Japanese interests participate. The Japanese share in Gregg River is 40 percent; in Bullmoose, 10 percent. The arrival of all this capacity on the world scene when coking coal markets were weak probably contributed to weakening world coal prices.

Australian producers were particularly upset at loss of sales to Quintette and Bullmoose. The Japanese have agreed to continue purchases from these two mines and pay prices higher than given other suppliers to compensate for the high mining and transportation costs. This is another example of the difficulties of long-term contracting. Response to market conditions different from those anticipated at the time of contract signing is difficult to effect. The party who is worse off than expected wants adjustment; the other party wishes to stick to its rights. Ultimately, a rearrangement is struck. The Australian reactions to Quintette and Bullmoose undoubtedly occur because these two mines endured less of an adjustment than others with contracts with Japan.

Colombia

The next important new exporter is Colombia. Exxon, in partnership with a Colombian government corporation, has built a large mine and the rail and port facilities to move the coal. Initial shipments from the project began in 1985 (Wright 1985, p. 69). The initial target is to move output from 6 million tonnes in 1986 to 15 million in 1989. By adding more equipment, capacity of 30 million tonnes and possibly even 40 million could be attained (Phillips 1984). The project is widely expected to be a formidable competitor because of low cash operating costs. However, doubts also exist about whether the $3,000 million invested will be recovered. World coal prices and how fast the operation can profitably reach high output levels are critical to profitability. In particular, rapid attainment of the full 30–40 million tonne output potential may be required to recoup fully the high outlays on transportation facilities.

Conclusions

Many participants have already emerged on the world market. The standard industry vision is of much capacity that is uncompetitive at 1985 prices. This invariably is attributed to the ineptitude of other companies (whom the discussants are eager to identify). The disinterested observer without access to cost data can only be sure that the discussions convey a picture that coal output has proved far easier to expand than anticipated and consumption has not grown as fast as expected. Prices and profits are at least below expectations. Even so, claims of unprofitability should be viewed skeptically.

However, this need not be settled. The information is clear enough to guide decision making. It indicates that considerable amounts of steam coal would be available into Europe at $50–$60 (in 1985 purchasing power) per tonne. It is difficult to imagine circumstances under which coal prices into Europe would regularly exceed $70 per tonne (compared to the $45 price in mid 1985). Thus, we seem to be in an era of ample coal supply. The tricky questions relate to exactly how the market will allocate supply and demand. The data are not good enough to answer this question, but suggestions can be made. This is done in the next chapter.

CHAPTER 9

Prospects for coal

In this chapter, the prior discussion is synthesised into an evaluation of coal prospects. While specific forecasts are noted, stress is on problems of appraising coal markets. Accurate quantitative estimations are neither feasible nor the critical aspect of a prediction. All we can do is assemble the available facts and indicate their consequences.

For that reason, multiple-case forecasts (scenarios) are preferable to a single estimate. It may be desirable to develop more cases than most multiscenario analysts usually do. Presentation of more extreme cases would better indicate the true underlying uncertainties. The benefits of forecasting are more in the exercise than in the numbers. The effort should force the analysts to consider what are the critical influences on the market and systematically appraise how they might interact to produce various patterns of production, consumption, and trade. These studies should be guided by recognition that the outcome takes place in a market and the principles of economic theory are indispensable to adequate analysis.

A further critical need is more explicit consideration of forecast-user requirements. While many are interested in the numbers for their own sake, others are seeking guidance for specific decisions. These decision-makers want to know what is the best choice to make, given present knowledge.

The prior suggestion that forecasters explore wider ranges of possible outcomes is one critical way to provide better guidance to decision-making. Another is to consider more explicitly what decisions must be made and what developments will have significant impacts on the payoff of the decision. Then forecasts could be designed to stress the critical factors. Decision-makers who lack time fully to read adequate appraisals need assistants who can do so. The critical points cannot be conveyed satisfactorily in one page summaries.

Decisions always are made on the basis of imperfect information. Forecasts are often wrong and unlikely to become better. Much that is relevant has not occurred and is inherently unknowable. It is difficult to know what will be invented and what uses they will find.

The turmoil in the world economy since the late 1960s has made it more difficult to isolate permanent from transitory changes. Many critical developments are specific to energy and particularly an exercise of monopoly power whose nature and persistence are uncertain. Therefore, energy prospects are especially hard to project.

Moreover, information costs money to obtain. Data gathering should be limited to that which provides aid equal in value to the cost of accumulation. Everyone has less data than could be gathered. A critical concern is that resource data are incomplete precisely because refinement costs far more than the information would be worth (recall Chapter 2).

The critical questions here are how much coal will be used in different countries for different end uses and from where will this coal come. This chapter begins with review of the economic principles that guide the development of future patterns. The material is used to analyse and provide alternatives to available studies of coal market prospects.

Stress is upon the OECD countries. They are simultaneously the most readily appraised and most interesting part of the world coal scene. The OECD nations constitute and will continue to comprise a large portion of the international coal and energy market. They absorbed about 79 percent of nonCommunist world coal use in 1984 and have a wide number of options, including major policy changes, open to them. Many studies are available on coal and energy in these countries. Moreover, review of OECD developments leads to consideration of several of the other critical aspects of coal, particularly the restructuring of world steel production and its implication for coal trade and the rise of new major coal exporting countries.

This leaves two hiatuses that are difficult to close – the Communist countries and the less developed countries, most notably India, with coal industries to meet internal needs. The prior discussion of the Communist countries included warnings of the great uncertainty about their coal prospects. Conditions in India, the largest producer not discussed in prior chapters, also are unclear.

Expectations about energy consumption growth and relative competitiveness have shifted radically since 1981 with drastic reductions in consumption growth projections. What is not clear is the extent to which the changes were caused by higher than expected oil price increases and to what extent by recognition that the ability to respond to higher prices by reducing energy use and raising alternative supplies had previously been underestimated.

A further concern is the role of politics and ideology in affecting forecasts. For example, the reluctance of public utility regulators in the United States to allow construction of new powerplants has created pressures for companies to lower growth forecasts.

Expectations about synthetic fuels shifted considerably. In the middle seventies, modest testing of the prospects was stressed. President James E. Carter's proposal for massive investments in synthetic fuels were taken at face value by forecasters. As support for his ideas diminished, so did predictions about coal use for synthetic fuels. As noted in Chapter 2, nuclear prospects differ radically among countries.

Another set of forecasting issues relates to the relative role of different countries (and regions within countries) in supplying coal. Most forecasters presume that protectionist coal policies will continue. Therefore, contractions will be limited to those announced.

This hiatus is a particularly graphic example of the pitfalls of reliance on government forecasts (or on "independent" estimates that all too often uncritically accept government assertions). The most accessible forecasts come from governments or international agencies sponsored by governments and reflect public policy. Although every past vow to maintain European coal mining capacity at prevailing levels has been broken, prognosticators are reluctant to guess what further retreats might be made.

Resolution of these questions requires that the analysts pay greater attention to both the development of coal supply in different countries and the public policies imposed. The effect of cost developments on public policy must be included. As the cost of protection mounts, pressures arise to cut output.

Economic analysis and coal market behavior

Prior chapters delineate the extensive differences among participants in the coal market. On the demand side, major end use and locational distinctions arise. Suppliers differ radically in their production costs, costs of delivery to key markets, and the quality of coal produced. The differences among users, in turn, involve both the nature and size of use. Coal is indispensable for traditional methods of ironmaking, feasible for any ordinary boiler, and unattractive for direct use in other energy devices (see Chapter 2). In any end use, large scale users are better able to use coal because the nonfuel cost disadvantage relative to oil and gas diminishes as the volume of heat needed increases.

As a result, many factors must be considered in an adequate analysis of coal market patterns. Coal usually competes with other fuels. Producers within the coal industry compete with each other. The competition is affected by the cost of devices to burn coal, transportation costs for different fuels from different locations, public policy, and the extent to which monopoly power is exercised in fuel markets. The more ample the supply of competitive fuels, the more pressure on coal. Thus, the worst case for coal is one in which ample supplies of oil and gas are available and nuclear power is cheap and politically acceptable. The less these conditions prevail, the better the outlook for coal.

Economic theory stresses that proper analysis involves consideration of the interaction among these influences. Rarely, if ever, can any element be considered to set either its price or its output independently of others. Energy analysts, nevertheless, often try to simplify their work by presuming that certain suppliers set prices independently of others' actions. Discussion supposedly only requires tracing the consequences of this preset price for world consumption and production patterns.

Such an approach has been used with the price of Middle Eastern crude oil, the world price of light or heavy fuel oil, or the price of US coal as a base. In every case, the procedure is dubious. A critical exception is Adelman's 1972 analysis of Middle East oil. He estimated the price that a competitive oil industry in the Middle East would charge. Implicit in his argument was belief that no reasonable change in his assumptions would cause competitive prices to lie significantly outside the range he projected.

At the time, he warned that his figures only showed the floor to world prices and that restraints on competition by consuming and producing country governments could lead to higher prices. (However, he indicated that the more likely development was continuation of the trend to slowly falling prices.) Subsequently, world oil prices moved far above his estimates of the minimum possible levels, and he was among the first to warn that the small rises of the 1971–73 period were the precursors of larger increases. The cause and consequences of these increases remain controversial.

Two basic, not necessarily mutually exclusive, explanations and many variants have been propounded to explain rising world oil prices. One view stresses exercise of monopoly power; the other, on growing scarcity of oil. Some observers blame developments all on one force; others see both contributing in various proportions. The increasing scarcity argument is precisely the one that Adelman's 1972 book tried to refute. At the time, many observers abandoned their belief that scarcity would soon lead to increased oil prices.

Subsequent developments should have reinforced rather than reduced belief in this argument. If Adelman was correct about production costs given his assumption of falling oil prices, then in an era of higher prices and the resultingly lower depletion of Middle Eastern oil, exhaustion of low-cost resources should have been delayed. Nevertheless, depletion fears eventually reemerged.

In any case, a model of increasing scarcity, monopoly, or a combination of the two involves the interaction of the Middle East with the rest of the world energy market.

Fuel-oil based analyses have been used as simple ways of studying the markets for coal and natural gas. Fuel-oil prices are deemed as setting a limit to coal and gas prices. At best, these arguments are the fixed crude-oil-price model in disguise. A fuel-oil price forecast implicitly involves assumptions about the price of crude oil. In addition, many further arguments must be made to relate that crude oil price to that of any one product.

This approach suffers from the defect of its most basic assumption, that fuel-oil prices are independent of coal and natural gas supplies. The fuel-oil price forecasts actually involve beliefs about coal and gas supply. In particular, many analysts and decision-makers expected natural gas to sell at parity with lighter fuel oils such as used in home heating. This implied that gas supplies would be too limited to allow continued penetration of the markets in which heavy fuel oil was an alternative. Similarly, a view that coal must adapt to a fixed price of heavy fuel oil gainsays any influence of coal and nuclear supply on oil prices.

A comparable argument is that US coal export prices are what will determine world coal prices. The rest of the world is presumed to adjust to a US price. Again, implicit theorising is being undertaken about the rivals. They are presumed unable to meet world demands at prices below those of US suppliers.

Such models are unsatisfactory because they are grounded in questionable appraisals of market conditions. Explicit consideration of interaction among suppliers cannot be avoided. The usual fixed price assumptions prove on examination to involve implicit views about other suppliers. Any price specified for fuel oils, for example, indicates some views about what end uses in what locations are likely to be served by oil. For reasons explained in Chapter 2, the price of a fuel must decline to expand the number of users.

In principle, a wide range of possibilities exists. Consideration of the available facts does not greatly reduce the number of plausible outcomes. Many visions of the future may arise from inadequate viewing and analysis of the data. However, it is not possible conclusively to prove this. Better analysis can only reduce the inherent uncertainties. Some observers are so committed to a position that no refutation will shake them.

The possibilities for coal extend from a shrinking, beleaguered industry to one with the return of dominance described in Chapter 1. The first outcome would arise if electric utilities shift to nuclear power and return to heavy use of oil and gas leaving the coal industry a static or even declining coke coal market. The other extreme would occur because it is necessary to use coal, not only for all boiler applications, but for manufacture of liquids and gases for those who cannot burn coal.

Many intermediate positions are possible, and most forecasts take a middle ground. The consensus prevailing in 1986 was that modest growth would prevail. Coal would maintain its position in electricity utilities and capture a significant part of the expansion in energy use in industry. The alternative I advocate is that while the forecasts for electric power are reasonable, gains in industry are far less likely (see Chapter 2).

Each forecast involves many assumptions about the market. One critical set of considerations relates to overall supply–demand conditions and the response of Middle Eastern oil producers to them. Several different forecasts of oil price behavior have been proposed.

One is a modification of the predictions prevalent in the 1970s of continued regular annual increases in oil prices. Given the declines in prices in the first half of the 1980s and the continued downward pressures on these prices, immediate price rises cannot plausibly be forecast. However, a forecast, popular in the 1950s and 1960s, that within a decade prices will recover as excess supplies disappear, has been revived. This might be termed the official view since it is so popular among officials of energy companies, governments, and international agencies concerned with energy.

The opposite argument is that the oil prices fall of 1986 is unlikely to be recovered for many decades. This might be termed the economists' argument. It reflects the belief that cartels cannot last. Many professional economists have expressed this view since the first oil price rises. The eventual collapse of all past cartels inspired similar expectations about oil.

In contrast to both positions, Adelman has long had ambivalent feelings about possible world oil price developments. He has warned that traditional doubts about the sustainability of cartels might not apply. He stresses the tension between monopolistic and competitive forces and that the power lies with governments rather than private companies. Government is better able to restrict energy production and use. Both producing and consuming country governments, in fact, favor use of these abilities.

While Adelman's earlier writings suggested that competition would ultimately reemerge, he has progressively moved away from that view. Shortly after his book appeared and before the 1973–74 price increases, he began warning that consuming countries would have to cease maintaining policies that encourage high energy prices. By the middle 1980s, he had become convinced that the prevailing forces will ensure the return of oil prices to monopoly-profit maximising levels. He proposes that the consuming nations make the best of this by taking the initiative from the producing countries. He advocates oil *import* taxes that would raise the price of oil to its monopoly level but with the receipts going to the importing countries. He implicitly is arguing the return will occur sufficiently soon that the short term losses to a tax will be offset by long term benefits of higher prices.

These various views arise from different opinions about both the market prospects and the nature of world oil price setting. The simplest rationale of the official forecast is the traditional depletion argument. However, more subtle arguments can be made that do not depend upon depletion of Middle East oil. It simply can be argued that a combination of resumption of vigorous economic growth and depletion of other energy resources will raise the demand for Middle East oil. Such an argument would stress that the slowdown in gross national product growth during the 1970s was largely the result of bad policies in the energy consuming countries, rather than to the oil price increases, or that the oil price increase impacts are behind us. Thus, growth can resume without being aborted by oil price rises. Additionally, the alternative energy supplies that held down demand for Middle East oil will become depleted.

Resort to models that do not depend on imminent exhaustion involves resolving questions about the most plausible form of price increase. One vision is that supply restriction is again effected in one big step and that subsequent changes in demand and the ability to produce are too modest to justify further price increases. The official model view of steady rises would prevail if price restricting power increased only gradually.

Similarly, the falling price model is most readily related to the expectations that cartelisation will collapse. However, one vigorous statement, by Robert Deam, bases the collapse prediction on the premise that cartelisation is simply monopolisation by Saudi Arabia. He argues that the Saudis severely misjudged what was optimal for them in the long run and set prices far above the level that would maximise profits on a sustained basis.

Deam believes the critical error was failure to anticipate the increased use of established alternatives for sharply reducing crude oil use. The oil industry has developed technologies that allow substantially increasing the proportion of gasoline, jet fuel, and lighter oils (sometimes termed white products) and reducing the proportion of heavy oils (or black products). These techniques were developed and heavily employed long before the oil price rise of the 1970s in the United States because ample supplies of natural gas and coal and high demand for gasoline made it attractive to increase the light product proportion. These techniques were less heavily used in Western Europe and Japan because of stronger relative demand for heavy oils. The rise in oil prices has altered this situation and encouraged substitution of natural gas, coal, and nuclear power for oil. This, in turn, has stimulated greater installation of facilities to increase the light product proporion of output. This has reduced crude oil demand more than expected.

As Deam also points out, the general failure to anticipate these demand effects increases the pressures on oil prices. Investments have been made in upgrading refineries, building natural gas pipelines in Europe, and installing coal-fired and nuclear plants around the world. Many of these facilities may prove bad investments should oil prices decline. However, continued operation of these facilities is cheaper than keeping them idle.

Views about oil and gas from non Middle East sources and world coal and nuclear power prospects critically affect their oil-price expectations. Those expecting low prices presume continued availability of such alternatives; those expecting price rises see the alternatives shrinking.

Different views on oil and on other influences can lead to the wide range of expectations about coal listed earlier in the chapter. The conclusions depend critically on more specific discussion than provided above of the relative competitive position of coal in the market.

Implications for coal

Key issues include (1) the degree of European coal output contraction, (2) the true competitive position of different coal suppliers, (3) how well coal can compete with other fuels, (4) growth in electric power, (5) world steel output, where it occurs, and the role coke will play, and (6) industrial boiler demand developments.

On the coal supply side, the position of the United States is a critical forecasting issue. At a sufficiently high price, the United States could supply very large volumes of coal for export. Whether sufficiently high demands and sufficiently low supplies of alternatives will allow the US actually to sell large export volumes is uncertain.

Such forecasts as the World Coal study and IEA work in the early eighties presented highly optimistic views of US coal. The critical premises were presumption of high levels of net demand for coal and the inability of others to cover these needs. Those involved with these studies now admit the analysis only showed what was attainable with extraordinary effort.

The analyses provided maximum figures on how much coal could be economically consumed. The ability to develop coal supplies in many different places received inadequate attention. The ebullient expectations about the United States followed inevitably from this approach. Effectively, the US was assigned all sales for which other suppliers could not be identified.

This optimism about US exports is now considered excessive. Neither what the proper estimate should be nor how lower protectionism in Europe would alter the situation is apparent. The cases discussed range from the US ending up as a net importer to that of substantial exports. Large exports require strong demands and a weak position of rivals. The net importer case would arise from a combination of inability to compete in the steam coal market in Europe, difficulties in coking coal sales, and the ability of foreigners to penetrate US electric power markets. The principal prospect is that substantial markets for Colombian coal could arise in the southeastern United States from Florida to Texas. However, it is doubtful that Colombian output would reach 50 million tonnes. Even if all this were sold in the US, the effects on what should be at least a 1,000 million tonne market would be modest.

Conceivably Australian and Canadian coal could find markets on the West Coast. However, as of 1986, no West Coast plants are well positioned to buy imports, and plans to add capacity that might have capability economically to use imported coal are moribund. Thus, US coal imports probably will remain much smaller than US coal exports.

The rest of this chapter attempts to provide indicators on these questions. The procedure is to start with various forecasts of energy market prospects and relate them to whatever is available on plans to meet these prospects. In preparing the discussion, numerous forecasts were examined and tabulated.[1] However, because of the obsolescence of earlier work, two 1985 studies, one from IEA and the other from Vincent Calarco of the Chase Manhattan Bank, are stressed. Then an evaluation is made.

Discussion begins with examination of the steel industry world wide. The effort is to suggest how much steel might be produced where, using what amounts of coal. Then the prospects for electric power and industrial fuel use in key OECD countries are reviewed.

Steel around the world[2]

The steel industry has undergone considerable changes involving the level and location of production and consumption, the technologies employed, and the sources of raw materials. The industry has moved from slow, steady growth to what is at best stagnation. Regional dispersion of production has increased. Over the very long term, the major developments have been the rise of the Soviet Union as by far the largest producer, the emergence of Japan as the second largest producer, and the depressed steel industry in the USA. Mostly since the early 1970s, South Korea, Brazil, and Taiwan have emerged as the most rapidly growing steel producing regions. Western Europe enjoyed steady growth but since the middle 1970s, it too has been beset by weakening markets.

These developments are associated with internationalisation of iron ore and steel trade. Japan and the three rapidly growing producers have become significant net exporters of steel. The penetration of these imports into the United States has been the most controversial aspect of these exports. However, the main outlets and the prime source of market growth since the middle 1970s have been less developed countries.

Local iron ore and coking coal have radically decreased their role in steel as large imports of ores and coal have become economic. The pattern of ore production has changed greatly since World War II. Some early entrants such as Canada and Venezuela have lost markets to newcomers such as Australia and Brazil. Relocation on the coastline to receive ore more cheaply became desirable. Another possibility is for the materials to be assembled in a third country such as Japan or Korea and produced for export. Ore rich countries such as Brazil might consider moving into the export of more processed products – pig iron, steel ingot, semifabricated steel, or finished products. The technological changes outlined in Chapter 2 also affected the market.

Ideally, forecasts of coking coal should explicitly consider how these critical influences will evolve. Few studies do this. An available coking coal forecast that does delineate its iron and steel forecasts, Calarco's, has a slowly declining trend for steel output in Western Europe, the United States, Canada, and Japan (see Table 9.1).[3] In the Japanese case, Calarco's full data show Japanese output rising from 1983 to 1985 and then declining.

Table 9.1 *Indicators of world coking coal consumption in 2000 (million tonnes)*

| | Steel production | | Pig iron production | | Coking coal consumption | | |
| | | | | | | Calarco | IEA |
	1983	2000	1983	2000	1983	2000	2000
Country							
United States	77	76	44	36	34	30	59
Canada	13	12	9	8	6	6	6
Western Europe	144	135	94	85	87	74	94
Japan	97	99	73	74	62	63	72
Australia/NZ	6	8	5	6	6	8	10
Total developed countries	336	330	224	210	194	181	241
Other Asia	30	59	21	41	36	53	NA
Latin America	29	62	23	35	16	34	NA
Africa/Middle East	12	28	8	13	8	13	NA
Total less developed countries	71	149	52	88	60	99	NA
Noncommunist world	407	479	277	298	254	280	NA
Communist Europe	211	199	132	118	195	148	NA
Communist Asia	46	98	39	69	55	98	NA
Total Communist	257	297	171	187	250	246	NA
World	664	776	447	485	504	526	NA
Ratios	NA	NA	0.67	0.63	0.54	0.51	NA

Sources: Vincent J. Calarco Jr., *World Coal Outlook: A Reassessment*
and International Energy Agency, *Coal Information 1985*, Paris, 1985
and *Energy Statistics 1982-83*, Paris, 1985.
Notes: The ratios in the pig iron column are tonnes of pig iron per tonnes of steel.
The ratios in the coke columns are tonnes of coke per tonne of pig iron.
IEA only reports OECD Western Europe; Calarco adds Yugoslavia.
IEA calculated as sum of changes calculated from *Coal Information* and levels calculated
'from *Energy Statistics*.'

The 1984 forecasts by Lenhard J. Holschuh of the International Iron and Steel Institute (IISI) indicate steel consumption by this group of countries will be level from 1985 to 1995. While the two estimates are difficult to compare, they appear roughly similar.[4]

Calarco predicts steel output by less developed countries will more than double from 1983 to 2000. Falling proportions of pig iron use per tonne of steel and of coking coal per tonne of pig iron also are expected. Coking coal consumption, therefore, drops more than steel output in developed countries and rises at a lesser rate than steel production in less developed countries (Table 9.1).

IEA's annual coal forecasts provide only indirect indicators of coking coal demands but seem much more optimistic than Calarco's.[5] Calarco differs most sharply with the 1985 IEA forecasts on the level of US coking coal consumption. He also sets lower levels for Western Europe than does the 1985 IEA study but gives similar numbers for Canada, Australia, and Japan (see Table 9.1). Calarco's numbers imply 2000 coking coal use in the US and Western Europe below 1983 levels but rises in Australia and Japan. IEA's 1985 projectives for 2000 involve a 25 million TCE rise in US coking coal use from 1983 levels and more modest gains elsewhere – about 7 million TCE in Europe and 8 million TCE in Japan.

Turning to the major exporters, the 1978 and 1985 IEA and Calarco net export forecasts for Canada prove quite similar. IEA's 1985 forecasts of US coking coal exports are well below its 1978 figures but above Calarco. Calarco and the 1985 IEA figures for Australia are similar and well below 1978 IEA estimates (see Table 9.2). Most forecasts thus suggest that Australia, the USA, and Canada will continue to dominate the coal export trade, but later works forecast lower growth than earlier ones. A range of 50–75 million tonnes for Australia, for the USA, 50–70 million, for Canada, 25–35 million tonnes prevails among all the forecasts examined.

Changes in European protectionism could alter the situation but probably not enough to increase greatly any other country's coal exports. In the mid 1980s, some 50–55 million tonnes of EEC coal went into coking – 30–35 million tonnes from Germany, about 9 million tonnes from Britain and around 3 million tonnes each from France and Belgium. About a third of the German output is exported as coking coal or coke; as Chapter 7 discusses, considerable criticism has been levied at the heavy subsidies required to effect these exports. Protection might keep this output stable. Otherwise, a 10 million tonne decline in Germany seems likely and halving of output in the other EEC countries could easily occur.

If Western European coking coal demand is over 100 million tonnes and European production vanished, European imports would then be more than double the levels forecast by Calarco and IEA. Given the differences in

Table 9.2 *Selected forecasts of coal consumption and output in different regions in 2000*

(IEA in mllion tonnes of coal equivalent ; Calarco in million physical tonnes)

Country	Total output	Net import	Use total	Steam output	Coking output	Steam imports	Coking imports	Electric	Coking	Other steam
Forecasts of OECD Europe										
IEA Steam Coal 1978	364	311	675	276	88	254	57	419	144	112
1985 IEA Coal Information	357	203	563	289	67	167	36	330	94	139
Calarco 1985	507	166	673	470	37	129	37	510	74	89
Forecasts of Japan										
IEA Steam Coal 1978	19	181	200	9	10	77	104	76	114	10
1985 IEA Coal Information	16	140	156	11	5	69	71	68	72	16
Calarco 1985	18	111	129	16	2	50	62	51	63	15
Forecasts of the USA										
IEA Steam Coal 1978	1,181	-129	1,052	1,019	162	-59	-70	800	92	161
1985 IEA Coal Information	1,250	-181	1,069	1,129	121	-119	-63	769	59	241
Calarco 1985	1,142	-81	1,061	1,060	82	-30	-51	933	30	94
Forecasts of Canada										
IEA Steam Coal 1978	71	-14	57	37	34	9	-23	45	11	1
1985 IEA Coal Information	103	-39	64	74	29	-17	-22	26	6	32
Calarco 1985	82	-29	53	56	26	-8	-21	44	6	3
Forecasts of Australia										
IEA Steam Coal 1978	285	-195	90	193	92	-120	-75	70	17	3
1985 IEA Coal Information	215	-128	86	152	63	+75	-53	63	10	14
Calarco 1985	239	-101	138	179	60	-49	-52	115	8	15

Source: International Energy Agency, *Steam Coal: Prospects to 2000,* Paris 1978.
      Vincent J. Calarco Jr., *World Coal Outlook: A Reassessment*
      and International Energy Agency, *Coal Information 1985,* Paris: 1985,
      Energy Balances 1982-83, and *Energy Statistics 1982-83,* Paris, 1985.
Notes: IEA only reports OECD Western Europe; Calarco adds Yugoslavia.
      IEA coking and other steam calculated as sum of changes calculated from *Coal Information*
      and levels calculated from *Energy Balances* and *Energy Statistics.*
      Calarco's use of physical tons primarily affects steam coal numbers.

total coal output among the three main exporters, the relative impacts of an additional 50–60 million tonnes in total sales would differ similarly. Even if all the gain went to the United States, it would be a negligible part of the 1,100 to 1,300 million TCE of output being predicted. Conversely, with Canadian output otherwise at 90 to 125 million TCE, a 60 million tonne gain would be substantial – probably more than could be supplied economically. With a 200–250 million TCE output prospect, Australia would be in a position intermediate between the US and Canada – with output being about 20 percent higher.

The logistics and economics suggest that in fact Australia would be the prime beneficiary of higher-than-forecast European coking-coal imports. However, if the overall increases are more in the 20–30 million tonne range consistent with Calarco's demand forecasts and a drastic cut in European coal output, the effects on Australia would be quite small. Stagnation in the coking coal market limits the ability of other coal producers to benefit greatly from reduction of European protection of coking coal.

Table 9.3 *IEA forecasts of coal use in the year 2000 (million tonnes of coal equivalent)*

	Amount						Change from 1983					
Region	Total output	Use total	Electric	Coking	Other steam	Nuclear	Total output	Use total	Electric	Coking	Other Steam	Nuclear
All OECD	1,945	1,942	1,257	241	444	790	757	700	463	46	191	514
North America	1,353	1,133	794	65	274	333	604	421	289	25	107	218
OECD Europe	357	563	330	94	139	327	37	168	96	7	64	206
EEC	251	408	264	73	71	266	-6	96	57	-1	41	173
Canada	103	64	26	6	32	60	56	18	-2	0	19	45
USA	1,250	1,069	769	59	241	273	547	403	290	25	88	174
Japan	16	156	68	72	16	130	1	67	44	8	15	90
Australia	215	86	63	10	14	0	113	41	32	4	5	0
New Zealand	5	5	2	0	3	0	2	2	2	0	1	0
Pacific	236	247	133	82	32	130	116	111	78	13	21	90
Austria	2	9	4	2	3	0	-1	1	2	0	-1	0
Belgium	6	22	12	8	2	16	0	9	6	1	1	8
Denmark	1	13	11	0	2	0	0	5	4	0	2	0
Finland	10	16	5	0	11	9	4	6	3	0	3	3
France	13	47	10	16	21	123	-5	7	-12	4	16	77
Germany	117	120	86	25	9	52	-8	-1	1	-6	5	31
Greece	14	22	19	0	3	0	9	15	14	0	2	0
Ireland	2	6	3	0	3	0	0	3	3	0	1	0
Italy	3	68	48	10	10	29	1	49	40	1	7	27
Luxembourg	0	3	0	0	2	0	0	1	0	0	1	0
Netherlands	0	13	8	4	0	7	0	5	4	1	1	6
Norway	2	3	0	0	3	0	0	1	0	0	1	0
Portugal	2	9	5	1	3	2	1	8	5	1	2	2
Spain	29	43	28	5	9	19	11	17	11	1	6	16
Sweden	9	16	4	2	10	18	2	6	3	0	3	5
Switzerland	1	2	1	0	1	8	0	0	1	0	-1	3
Turkey	50	58	19	11	28	5	26	33	15	7	11	5
United Kingdom	94	94	66	10	18	40	-3	2	-2	-2	6	24

Source: Computed by R.L. Gordon from International Energy Agency, *Coal Information 1985*, Paris 1985, and Organisation for Economic Co-operation and Development, *Energy Balances of OECD Countries*, Paris:1985. See text for explanation of computation.

Non-utility steam coal

Appraisal of this sector is hindered by the demand uncertainties noted in Chapter 2. As explained in footnotes, the methods used to infer the forecast can produce large errors. Calarco's steam coal forecasts are hard to compare with other projections because he uses physical tonnes rather than TCE.

IEA projects the biggest gain in the USA which will continue to account for the majority of other steam coal use in the OECD. The total gains in OECD Europe are predicted to be less than that of the USA. Moreover, the French rise is far greater than that elsewhere – roughly similar to that expected in Japan (see Table 9.3).

The justification for these forecasts is that they can be attained by a few replacements by coal of oil or gas fired boilers in large plants. The offsets include both the oil and gas price and environmental problems discussed above and a tendency for coal use currently to occur in contracting industries. Again, the amounts are small particularly if shared by several suppliers.

Coal in electricity generation

As noted in prior chapters, three considerations are critical about electric power and coal. First, electric power is the principal and most clearly expanding market for coal. Second, the future of electricity growth is unclear. Third, to various degrees, the use of coal is threatened by alternative energy sources. The first point is covered in earlier chapters; stress here is on the last two.

The electricity prospect side of the question is an example of the forecasting uncertainty problem discussed at the start of the chapter. The only feasible elaborations are reviews of the different forecasts. This seems unnecessary for present purposes. More attention should be given to the problem of alternatives to coal.

Stress usually is on nuclear power, but it is one consideration. Indirect resort to nuclear power may emerge as an alternative in both the United States and western Europe. Spare nuclear capacity in Canada and France has already inspired electricity exports, and they may increase significantly. The United States also may move towards imported power generated from other prime movers and resort to small scale, noncoal burning alternatives. Finally, oil and gas price trends may be such that moves to replace those fuels with coal may not proceed further.

Those OECD countries with significant thermal electric industries – the United States, Canada, Japan, Australia, Britain, Germany, France, Italy, and Spain – are emphasised. What these countries do will be the major influence on free world coal consumption.

Australia is the most straightforward case. Its (economic and political) policy of reliance on coal means that coal developments depend almost entirely on the rate of generation growth.

The Japanese case involves additional questions of what adjustments will be made in the effort to diversify energy sources by using a mix of nuclear, coal, and liquified natural gas. IEA's 1985 report estimated year 2000 Japanese coal use in electricity generation at 68 million TCE while nuclear would displace 130 million TCE. Calarco, with an almost identical forecast of total generation, only sets coal use at 50 million (physical) tonnes. Changes within this range would only modestly affect world trade.

Questions arise about the role of power imports and the mix of plants that actually will be completed in the US. US public utility regulation directly and indirectly discourages the construction of the large scale plants that traditionally have been the basis of the electric power industry.

New capacity is unattractive to regulators because of the substantial electricity price increases that are necessitated. The plants have higher real costs than those previously built. Moreover, electricity prices are based on the historical current dollar costs and held down by the large portion of facilities built before the severe inflation of the 1970s. Adding new plants then reduces the weight of old facilities in rate determination and further increases rates. Proposed construction, completion of plants under construction, and even allowing rates to reflect costs for completed plants have been explicitly prohibited in some states. Limits on rates of return further discourage construction of new facilities. These problems are most severe for nuclear plants because US government nuclear regulations supplement the pressure exerted by the states.

Coal plants are also affected by these pressures. California initially made new nuclear construction contingent on certification that a "solution" for nuclear waste storage has been developed. Then the state discouraged construction of coal fired plants in the state or even participation in more of the out-of-state coal ventures to which California utilities had previously moved. California utilities are stressing small units using cogeneration, solar, or biomass in their expansion plans (see North American Electric Reliability Council, 1985). Texas is among other states that have begun efforts to encourage alternative energy sources. In addition, California and states on the Canadian border are looking to imports of power from Canada.

Navarro (1985) warns that the pressures to reduce expansion have gone too far and that the need for new supplies will not be recognised soon enough to institute and complete projects for coal fired plants. Instead, quicker-to-complete oil or gas using units will be built. This last, however, seems to ignore that the completion times could be greatly cut if the need were recognised.

All this adds up to great uncertainty about the level of coal use in the United States. What is more certain is that most of the growth will be in the South and the West. Therefore, regional coal production growth patterns are likely to be similar to those that prevailed from 1970 to 1985.

France with its strong commitment to nuclear power has the most clearly delineated electricity situation in Western Europe. Basically, Electricité de France is likely to cut coal generation, but Charbonnages de France will make a greater effort to maintain its generation. It too will be under pressure of cost reduction efforts.

Britain's largest utility – the Central Electricity Generating Board (CEGB) serving England and Wales – had eight nuclear and two coal units (all in the 600 megawatt range) under construction in 1985. All but two of the nuclear units were scheduled for completion by 1986. The South of Scotland Electricity Board's expansions are all nuclear. As Chapter 7 indicates, the failure to complete an inquiry into British adoption of light water reactors has delayed planning new CEGB plants. The evidence suggests that coal use by British electric utilities will rise modestly if at all from its 1983 levels.

Italy only began in the 1970s to move away from heavy dependence on oil. The biggest change was an increase in coal use; natural gas consumption also rose. Italy's principal utility, Ente Nazionale per L'Energia Elettrica (ENEL), has a program for about 12,000 megawatts (MW) of coal capacity and 13,000 MW of nuclear. The coal additions supposedly would be completed by the early 1990s; the nuclear, a bit later. However, considerable barriers to completion seem to exist.

As Chapter 7 noted, West Germany may have problems of simultaneously complying with the contract of the century, using as much low cost lignite as desired, utilising nuclear capacity at preferred levels, maintaining, let alone increasing, coal imports, and using oil and gas when economic. Thus, the 1985 IEA forecasts for 2000 set total solid fuel use in electricity generation at 86 million TCE. If lignite use stayed at the 30 million TCE level that has prevailed in the 1980s, about 56 million TCE of hard coal would be used. This forecast suggests limited scope for coal imports if the 46 million (physical) tonne contract obligation remained in force.

The import potential implicitly in these estimates is about 10 million TCE. The errors in estimates of total electric utility energy use and of the amounts from different sources are each subject to at least a 10 million TCE error. Much different actual coal import patterns could develop. Lower generation and a few more nuclear units would suffice to create all the feared difficulties. Conversely, a bit higher output level, lags in nuclear completions, and lessened lignite use could allow absorption of the contract of the century amounts. However, the high costs of that accord would remain.

Overall, the 1985 IEA report indicated electric utility coal use in 2000 will be 96 million TCE above 1983. The biggest gain by far is in Italy – 40 million TCE. Again the rest is made of numerous scattered increases (see Table 9.3).

All these changes are dwarfed by the projected gains for the United States. IEA reports them at 290 million TCE; Calarco has a rise of 450 actual tonnes. Given the high proportion of output increases accounted for by lignite and subbituminous, his number probably differs less than it might seem with IEA. Thus, again, the gains now expected in the markets involved in world trade are modest.

The overall prospects

At least four problems remain in dealing with these forecasts. First, I have not yet considered whether the modest parts sum to a significant total. Next is uncertainty about whether the figures are still plagued with overoptimism or a trend to excess correction emerged. Then, the impacts of lesser protectionism must be considered. Finally, that sales might be allocated among suppliers should be appraised.

The prior section has suggested that coking coal consumption gains, if any, in OECD Europe and Japan would be modest. Increases in nonutility steam coal would be about 50–70 million TCE in Europe and 10–15 million TCE in Japan. The gains in the utility sector would be 70–90 million TCE in Europe and 30 to 50 million TCE in Japan. Thus, the 1985 forecasts examined imply increases of European coal use in the 125 to 150 million TCE range. The levels of 550 to 575 million TCE can be contrasted to earlier estimates of 675–800 million TCE. A 50 to 75 million TCE gain is set for Japan while earlier estimates implied gains up to 150 million tonnes. In the newest forecasts considered, the coal importing regions of OECD are expected to raise coal use by 175–225 million TCE compared to earlier estimates of gains in excess of 500 million TCE.

The 1985 forecasts imply modest gains that if evenly distributed among suppliers would not provide significant unexpected stimulus for anyone. The scaling back has eliminated the glowing growth prospects previously projected. In fact, the realisation of 1985 IEA forecasts for export changes for the US, Canada, and Australia probably would require the rise of significant non OECD markets.

A total export gain between 1983 and 2000 of 228 million TCE is set, for the USA (113 million TCE), Australia (78 million TCE), and Canada (37 million TCE). However, the predicted gain in OECD Europe (128 million TCE) and Japan (67 million TCE) is only 195 million TCE. Moreover, South Africa and Colombia also will raise exports. Calarco sets gains of about 50 million tonnes for South Africa and 31 million tonnes for Colombia. If this is added to the IEA forecasts for the other leading exporters, a total gain of 309 million TCE is implied – 114 million more than the rise in OECD Europe and Japan.

Calarco's forecasts for exports from the USA, Australia, and Canada and his import forecasts for OECD Europe and Japan are below the 1985 IEA estimates (Table 9.2). He also provides estimates for the non-OECD countries suggesting that non-OECD markets are unlikely to grow as much as is necessary to realise the 1985 IEA forecasts. He sees a rise of around 70 million physical tonnes of coal imports by less developed countries and little rise in Communist block imports.

Table 9.4 *Alternative forecasts of year 2000 coal output in OECD Europe*

(million tonnes of coal equivalent)

| | Forecast 2000 | Actual 1983 | Change | Adjusted cases | |
				Low cutback	High cutback
United Kingdom	94	98	-3	95	60
Germany	117	125	-8	55	10
Turkey	50	24	26	25	15
Spain	29	19	11	14	9
Total above	291	265	26	189	94
Greece	14	6	9	6	6
France	13	18	-5	13	5
Finland	10	6	4	6	5
Sweden	9	7	2	6	5
Belgium	6	6	0	6	0
Italy	3	2	1	2	2
Austria	2	3	-1	2	2
Ireland	2	2	0	2	2
Portugal	2	1	1	1	1
Norway	2	2	0	2	2
Switzerland	1	1	0	1	1
Denmark	1	1	0	1	1
Total Other	66	55	11	48	32
Total OECD Europe	357	320	37	237	126

Source: First 2000 forecast and 1983 actual from International Energy Agency,
　　　Coal Information 1985, Paris, 1985.
　　　Last two columns are R.L. Gordon estimates of the range possible
　　　with reduced protection.

Among the developments that might make figures higher are better than expected performance by the steel industry, difficulties for nuclear power, and another oil price rise. The main things that could hold back growth are slower market expansion, decisions to go more heavily nuclear – either by building plants in key countries or by imports elsewhere in Europe of electricity from France, and further drops in oil prices.

At one extreme, growth in coal use in OECD Europe and Japan might total less than 100 million TCE if the only gains were limited ones in electric power. Conversely, displacing 100 million TCE of nuclear power and gaining another 100 million TCE through better than expected market growth is also possible. Thus, the plausible range expands to imply market growth of between 100 to 450 million TCE from 1983 to 2000.

The impacts of reduced protectionism also are uncertain. Two countries, Germany and the UK, account for the majority of 1983 and projected 2000 coal output in OECD Europe. The next most important countries are Spain and Turkey. The rest of output in OECD Europe is scattered in several countries with limited growth prospects (see Table 9.4).

Available information indicates UK mines have a much better competitive position than the German. Compilations of mine-by-mine costs published in a 1983 Monopoly and Merger report on coal suggest at least 55 million tonnes of extant underground capacity is competitive. Another 25 million TCE might be producible from low cost open cast mines; planned new mines could add another 15 million TCE. Thus, the IEA UK forecasts might be attainable without protection.

Predicted German production is subdivided between 38 million TCE of lignite and 78 million TCE of hard coal. The standard forecast is that cheap surface mining of lignite can continue indefinitely at this scale but that probably very little, perhaps 10–20 million TCE, of hard coal could compete. Optimism about lignite should be qualified because controls on both its production and use may increase greatly. Land disturbance has become more controversial. The stringency of air pollution regulations in Germany may make lignite too expensive to use. Flue gas desulfurisation may become necessary. Thus, instead of 55 million TCE of solid fuel surviving, output might fall to 10–20 million TCE.

IEA discussions of energy and coal policy regularly indicate problems of actually assembling the resources to effect expansion of Turkish lignite output. However, much of the difficulty is alleged to be institutional. A nationalised company with limits on the salaries it pays had exclusive control of lignite mining. Private company participation is being facilitated. Nevertheless, Turkish output could actually stay close to its 1983 level of 24 million TCE.

The 1985 IEA forecast for Spain implies a rise of about 11 million TCE above 1983. Actual output, given lower protection, could be 5 to 10 million TCE lower than 1983. It is even more difficult to guess how much protection prevails elsewhere. Nevertheless, without protection solid fuel output is likely to be at least 100 TCE below the 1985 IEA forecasts for 2000. One way this could occur would be with the UK attaining forecast levels, Germany only to produce 55 million TCE, no growth in Turkey, and Spain falling below 1983 levels.

If German output were to fall to 10 million TCE and Turkey and Spain produced significantly less than the 1983 levels, a further 65 million TCE decline is implied. An equal additional decline could result from various combinations of British inabilities to produce at the forecast levels without subvention and substantial drops in countries not explicitly reviewed (see Table 9.4). Thus, it is possible a fall of as much as 230 million TCE might occur. In addition, a few more million TCE would result from ending Japan protectionism. If this all was displaced by imported coal rather than by a mix of fuels, it would double coal imports.

Combining the 100–400 million TCE range for consumption growth with a 100–240 million TCE estimate for production declines means import growth in the OECD could range from 200 to 640 million TCE. In contrast, a 400 million TCE consumption rise is predicted for the USA.

The opinion of observers of the world coal market is that a world trade expansion similar to that forecast by Calarco – i.e., in the 150–200 million tonne range – can be effected without noticeable price rises. A rise in the 350–400 million tonne range suggested by the IEA forecasts might (or might not) require recourse to Appalachian US capacity requiring $5–10 higher prices per tonne. It is only at the upper levels of my import quantity range that price uncertainties emerge.

Just what it would cost to produce exports in excess of 400 million tonnes has not been adequately considered. Even here one can imagine another 50–100 million tonnes of capacity in the five countries being considered and perhaps 50–100 million more from countries not yet considered being produced without further cost rises.

The discussion thus far deliberately has ignored the effect of price on consumption. The high figure for coal trade probably is attainable only if very ample supplies of coal are available at not much higher prices than prevailed in 1985 (and other conditions are also favorable). Thus, what really is available greatly affects the outcome.

Finally, the visions reported here suggest that the (best identified) lowest cost portion of the world coal supply curve consists of the best Australian and Canadian mines, the operating mines in South Africa, and the new operation in Colombia. More output from each of these areas and a great deal from the United States could be added if prices rose $5–10. The Calarco and IEA forecasts are reflections of these views. Calarco has lower exports for the US because he has a lower demand than IEA and thus implicitly a price that keeps US coal from competing. At the higher IEA forecast level, the surest way to meet the increases is to offer higher prices to US producers. The open question is how much unexploited low cost resources exist elsewhere in the world to put a further limit to the price impacts of expanded world trade. Those who invest in this realm clearly recognise that much uncertainty exists and flexible responses must be facilitated.

Further complications are created by exchange rate uncertainties and political instability in South Africa. In short, once we get past recognising that coal consumption is likely to grow, it is difficult to provide precise estimates of the location and end use of consumption or of who will supply it.

CHAPTER 10

Summary and conclusions

This study provides an overview of coal in world energy markets. The basic premise is that coal is subject to economic principles governing all other commodities and prospects depend upon the economic availability of coal resources.

The larger magnitudes of coal resources are irrelevant. The question is how much coal can be sold profitably. The physical abundance advantage is possibly an illusion; the true physical availabilities of alternative fuels are unclear. The critical consideration, however, is how much of the coal supply can be mined cheaply enough to offset the substantially higher transportation and capital and nonfuel operating costs in utilisation compared to those for oil and gas. Coal use is most economic in large scale facilities such as electric power plants because cost disadvantages narrow enough to make coal use more attractive. Cokable coal is indispensable to iron-making. However, this market has been depressed since the late 1970s. A combination of reduced demand for steel and technical developments that reduce the use of coking coal per tonne of final output of steel have held down demand.

The world's cheapest to mine coals can successfully compete in electric power stations around the world. Coal from such sources has been the principal fuel for electricity in the country of production and some other countries. Growth of that market has helped coal sales rise. Coal then, while still less widely used than hydrocarbons, enjoys steady growth. The principal challenge is that something better such as nuclear power may come along to disrupt this pattern.

A dozen countries account for over 90 percent of world coal output (Table 4.1). They differ radically, however, in size and strength. The United States is the leading producer primarily because costs are sufficiently low to make it economic to generate the majority of US electric energy from coal. It also competes in the coking coal export market and serves the depressed US coking coal market.

On a world scale, the much smaller Australian and South African industries have undeveloped capacity with lower costs than coal from new mines in the United States. Thus, Australia and South Africa have developed substantial export markets. The Australians started with heavy emphasis on coking coal and subsequently moved into substantial steam coal exports. Exports exceed domestic consumption. South African exports began with small coking coal sales but quickly moved to predominant stress on steam coal. Canada also has developed a significant position in exports, again of coking coal.

The Communist block and particularly China and the Soviet Union, respectively the number two and number three world producers, have more uncertain coal economics. Soviet coal production growth ceased in the late 1970s. The Soviet Union, China, and the Eastern European CMEA countries face difficult energy decisions. Western specialists on the subject face great problems of appraising the situation. The lack of a rational price system and the limited data availability hinder appraisal. The historic importance of coal to these countries and the desire to increase coal use are clear, but the economic and political realities that will determine future developments are uncertain. The true economics of the choices are unknown, probably by the Communist country planners as well as by Western observers. Similar questions arise about what political obstacles exist and how willing the policymakers will be to overcome the difficulties. Whether coal expansion is either desirable or attainable is quite unclear.

Other countries such as Japan and most coal producing states in Western Europe have clearly uneconomic coal industries. Their existence is rationalised as contributing to such nonmarket objectives as less disruption of employment and security of supply. This study argues that policies cheaper than coal protection are available to meet these policy goals.

Whether coal industries thrive or langish, governments intervene. At the least, action is needed to control the environmental impacts of using coal. States also believe that regulation of the environmental effects of surface mining is desirable and pass health and safety rules on the premise that workers are incapable of adequately protecting themselves from workplace hazards.

Industries perceived to be prosperous will be heavily taxed. These taxes can reach levels that adversely affect production and still prove difficult to reform. The US Congress has aggravated rather than alleviated coal leasing problems; Australia is making efforts to lessen tax burdens on coal.

A declining coal industry inspires a wide variety of government aid as is vividly shown by Western European experience. Assistance has been given for worker relocation. Critical debts, principally pension obligations, of the industry have been taken over by governments. Many government policies are used to prop up the coal industry. These include classic measures such as import control and many forms of subsidies. More unusual arrangements have developed to force coal purchases. West Germany imposes a tax on electricity sales to subsidise some electric utility coal use. The bulk of sales, however, occur under government ordered contracts between the coal mining companies and the individual utilities.

While continued growth of coal use seems likely worldwide, the most certain and important parts of the growth will occur in the United States and China. Japan and Eastern Europe will continue to increase coal use but probably remain smaller consumers than the US and Western Europe. Western European prospects are quite unclear, but substantial coal-consumption growth is unlikely. Radical policy reform might double Western European coal imports. However, when spread among several suppliers, the rises might not make a major contribution to increasing sales in any one country.

Coal appraisal then discloses numerous complexities. The industry is growing but not to preeminence. The growth is concentrated in a few strong industries. Other countries have coal production surviving only through massive government subsidies. These countries have allowed their coal industries to contract, but only at a painfully slow rate. Thus, the desirable changes worldwide involve further contractions as well as gains.

Given the particularly poor record over many decades of optimists about coal and the widespread errors most energy analysts including me have made since the early 1970s, the prudent position is to caution that the details of coal development are highly uncertain. This should be accepted as the unavoidable consequence of prevailing knowledge. Moreover, nothing critically depends upon knowing the answers. The world can afford to wait for the events to develop.

Notes

Chapter 1: Introduction – a perspective on coal

1. Many books have appeared over the years dealing in various degrees of detail with the nature of coal formation and with the technology of coal use. See, for example, Grainger and Gibson (1981), Speight (1983), and Merrick (1984). While these books seem to cover the subject adequately, they are simply the first 1980s works of which I became aware. Less comprehensive but much more comprehensible discussions appear in Schmidt (1979).

Chapter 2: An introduction to coal in the energy market

1. These definitions deliberately neglect the many finer subdivisions employed and how they differ among countries. The only distinction needing note is that outside the United States both bituminous and anthracite are called hard coal; the US employs the term for anthracite. The broader definition is used here.
2. The practice of using tonne to denote metric tonnes is employed here. Throughout, an effort is made to distinguish among the different measuring rods used and particularly the distinction between the tonnes of solid fuel actually produced (called physical tonnes when ambiguity might arise) and their equivalent in terms of coals of a standard quality. This latter concept is explained in Chapter 3.
3. An alternative method of securing access from the surface is to use augers (roughly giant drill bits) to extract coal particularly from seams on hillsides. The method is used in the US, mostly in Appalachia and, in fact, predominantly in eastern Kentucky and West Virginia.
4. This hiatus reflects the absence of incentives to undertake elaborate surveys of the exact uses of energy. Data collected from energy producers provide most of what we now possess; extensive material from consumers would be needed to obtain details on how the energy was used. Need for these data has not been sufficient to justify the expense of collecting them. The US did collect fuel use data by sectors of the manufacturing realm and report them in the *Surveys* and *Censuses of Manufacturers*.
5. A tabulation of thermal efficiency used to be included in US Department of Energy, *Thermal-Electric Plant Construction Cost and Annual Production Expenses*, and its predecessors. The last year covered was 1977. British experience is somewhat different, probably because oil plants in Britain are more modern than British coal plants.
6. Among the better accessible dispassionate discussions of the nuclear debate are in Keeney, 1977, and Landsberg *et al.* 1979. Evans and Hope, 1984, well survey nuclear prospects in different countries.
7. Predictions of such rises were frequently made by a wide variety of coal forecasters including the International Energy Agency and the US Department of Energy. Numerous cost calculations have been made purporting to prove the shifts are profitable. A particularly interesting series of studies of the UK have been made by Skea, who has noted many of the difficulties noted here.

8. More than a half century after its publication, Irving Fisher's *Theory of Interest* remains the best discussion of the subject. A more modern, more difficult treatment is provided by Hirschleifer (1970). Among the popular texts are Bierman and Smidt (1975) and Merrett and Sykes (1973).

Chapter 3: Coal consumption trends in OECD countries

1. This overview was developed by analysis of data in tonnes of oil equivalent appearing in the *British Petroleum Statistical Review of World Energy*.
2. Numerous tabulations were made to facilitate analyses of coal use in the OECD countries. Basic energy balance data for 1960–83 were tabulated for OECD Europe as a whole, Britain, Germany, Italy, Japan, and Australia. A separate tabulation of US data already existed. These countries are far larger energy users than any others. A tabulation of coal consumption data from 1950 to 1983 added Belgium, the Netherlands, and Spain because of their large coal use.

    Balance sheet solid fuel data, and the coal and lignite tonnage data were compiled for all OECD countries in 1982 and then 1983. Tables on lignite from 1950 to 1963 were made for the countries in which it is important – Germany, Australia, the United States, Canada, Greece, and Turkey.
3. The other OECD countries not listed are Austria, Norway, Sweden, Switzerland, and Turkey.
4. The six original members of the Communities were Belgium, France, Germany, Italy, Luxembourg, and the Netherlands. Britain, Ireland, and Denmark joined in 1973 and Greece in 1981. Spain and Portugal became members in January 1986.
5. By the nature of the measuring rods, these can be presented in many different equivalent ways by varying the amount of fuel valued or the heating value unit. Thus, the oil equivalent is also 10^7 kilocalories per metric tonne, 10,000 kilocalories per kilogram, or 10,000 teracalories per million metric tonnes. Also, the variation of heat contents of coal is considerable. In OECD data, nonlignitic coal is valued at a low of 5.8×10^6 kilocalories per tonne for New Zealand and Spain and a high of 8.2×10^6 for Finland. Thus, actual tonnes of even bituminous and anthracite may exceed their TCE level. The norm is good, rather than average, quality coal.

Chapter 4: Trends in coal production and trade

1. The Ruhr coal mine owners' association (Unternehmensverband Ruhrbergbau) published a series of coal data compendiums that provided figures from 1900 to the late 1950s. Every year from 1920 was reported for all major producing countries. Full 1900 to 1919 data were available for some countries; for others only 1900, 1905, 1910, and 1913 were reported. Lignite data, however, only go back to the 1920s. A table of coal output for leading countries appears in reports of Statistik der Kohlenwirtschaft.
2. The historic data are in the Ruhr coal owners' report cited above. More recent data that attempt to be inclusive appear from both the German coal statistics bureau (Statistik der Kohlenwirtschaft) and the US Department of Energy. The former's figures were used because they cover more years and are in TCE rather than Btu.
3. This section is based on coal data reports from each country. The US is from government reports; the British and French from the nationalised coal industries' reports; the Belgium from reports by industry associations.
4. Although this region is Flemish and the company issues its official annual reports in Flemish, the French spelling is normally employed in English language discussions and is used here. The Flemish spelling is Kempen.
5. Historically, these operations occurred on both sides of the border between Lower Saxony and North Rhine Westphalia, but the region was called Lower Saxony. After 1978, the closing of the Lower Saxony part of the operation caused the region to be renamed Ibbenbüren, the town in North Rhine Westphalia near which the operations occur.

Chapter 5: Coal in Communist countries

1. The BP data used for other countries were compared to a 1985 report on China issued by the World Bank. The BP numbers show a higher coal equivalent tonnage and coal share than does the World Bank. The numbers in the text are consistent with both estimates. The key seems to be differences in estimates of the heat content of lignite. Comparisons are possible with production data. BP shows the actual tonnages of hard coal and lignite output and their oil equivalent. The World Bank also gives the tonnages and their hard coal equivalent. Comparison of the data shows the main difference is in the coal equivalents computed from similar estimates of the actual tonnes.

Chapter 6: Coal in the USA – the public policy issues

1. Samuelson (1954, 1955, 1958, 1967, 1968, 1969) wrote the classic modern discussions of the problem. Many others have commented and discussed how the question can be resolved. Key contributions include Coase's 1960 warning that the defects of government decision making are a major impediment and work such as Sugden 1984, discussing how voluntary subscription systems might be designed to produce more efficient results than earlier writers such as Samuelson expected.

Chapter 7: The state and coal – the Western European case

1. This discussion combines material in Lister 1960, the Mining *Jahrbuch* published by Verlag Glückauf, and the annual review of Statistik der Kohlenwirtschaft.
2. The tax rate information was supplied by the Economics Ministry of the West German government.
3. The material for the rest of the chapter comes from Gordon (1970), a 1977 EEC report on coal, other EEC reports, the IEA surveys of coal and energy policy, various reports of the nation involved, energy reports of the Economics Ministry of North Rhine Westphalia, and discussions with European sources.
4. The European Currency Unit was established in 1976 to provide a new common denominator for different monies. The most accessible explanation and listing of historic data appears in the EEC annual, *Basic Statistics of the Community*. It is a weighted average of EEC member currency values from 1969 to 1973. Gross national products and trade volumes determine the weights. The unit replaced the European Unit of Account, which was the US dollar disguised as its gold value.
5. The principal justification for omission is the difficulty of dealing with the complexities of many different national policies and the efforts to reach international accord. In addition, my knowledge is limited.

Chapter 8: The emergence of producers for export

1. This section draws heavily on data from *Black Coal in Australia* and the Queensland Coal Board's annual report. The reliance here on fiscal year data is because, in many cases, more fiscal years than calendar years are covered in the reports. Similarly, some data are shown on an uncleaned tonnage basis because the cleaned coal figures are not reported. *Black Coal* provided the production data for all of Australia and mine output and ownership for NSW. The Queensland report gives mine output and ownership for Queensland. Donald Barnett has prepared several useful surveys of Australian coal.
2. This discussion of South Africa comes from interview material during a 1983 visit and a variety of annual reports secured subsequently. These cover Anglo American, Amcoal, Gencor and its coal subsidiaries, Barlow Rand and its coal subsidiaries, Johannesburg Consolidated Investment (JCI), Sasol, and Escom. In addition, both the statistical reports of the Chamber of Mines and annual reports of the Department of Mineral and Energy Affairs were secured. A major problem has been securing reports. In particular, due to some technicality precluding their mailing to the US, Barlow Rand material has been

secured only through 1982. The discussion has been pieced together from all these sources. The Chamber of Mines reports were used as a starting point. A list labelled sales by member companies is provided. However, the coverage is incomplete. Only 64 percent of 1984 sales were included. Most of the omissions relate to omission of mines wholly owned by members. Specifically, although Sasol is a member and reports coal output in its annual reports, its use is not reported by the Chamber. Another omission is a large Barlow Rand mine serving Escom and selling Escom almost as much as the reported Barlow Rand output. These omissions account for about a quarter of total sales. Another 10 percent of output then remains omitted. Perhaps 10 million tonnes or six percent is represented by the mines Barlow Rand operates for Shell and BP. A further complication is that many companies report on fiscal years different from the calendar year.

3. The most frequently occurring pattern was grants of 300,000 tonne quotas divided equally between high and low quality coal to seven companies. Three companies were allowed 300,000 tonnes of high quality coal exports and 150,000 tonnes of low quality coal. Two companies were given 200,000 tonnes of new high quality coal exports and the right to transfer exports to Richards Bay. Each of the other twelve allocations was unique. Four were solely for low quality coal and ranged from 50,000 tonnes to 750,000 tonnes. One company only received the right to relocate 240,000 tonnes to Richards Bay. Four companies received from 800,000 to 1,050,000 tonnes of quotas spread among all three categories. Two other companies obtained *only* quotas for new high quality coal exports and transfers to Richards Bay. In one case, the mix was 260,000 tonnes new and 240,000 tonnes transferred; in the other, 500,000 and 180,000 tonnes. One company only received 500,000 tonnes in new high quality coal quotas.

4. This review of Canada stresses exporting firms. Basic output data came from Keystone, world trade data from National Coal Association and German reports, and the current history from the *Canadian Mining Journal*.

Chapter 9: Prospects for coal

1. The studies viewed include several IEA forecasts from 1978 to 1985, the World Coal study, two EEC forecasts, one from the Economic Commission for Europe, two from Saarberg-werke, and numerous forecasts of the US from the Department of Energy and private consulting services. The advantage of the IEA forecast chosen is that it provides particularly detailed data on every OECD country. Calarco's forecast has worldwide coverage and shows great care in specifying its basis. It also represents a sharply different view from IEA.

2. For the purposes of this section, data from the International Iron and Steel Institute and American Iron and Steel Institute's statistical reports were employed.

3. The appraisal is complicated because he uses 1983, a depressed year, as the historic year for comparison. His forecasts then show a recovery of output in 1985 and then declines. For example, the US is projected to go from 77 million in 1983 to 82 million tonnes in 1985 and 76 million in 2000. By 2000, Canada is predicted to be at about its 1983 level of 12.5 million tonnes; Western Europe goes from 144 million in 1983 to 135 million in 2000; Japan from 97 million to 99 million.

4. By definition, production equals consumption plus net exports plus inventory change. The last is thought to be negligible in such stagnant markets. Thus, if net exports are level, production and consumption *changes* would be identical. IISI data indicate that for the industrialised countries as a whole, net exports were stable at around 40 million tonnes from 1973 to 1983. More detailed comparisons between IISI and Calarco involve resolving differences in the regional breakdowns each use. IISI's sectors are North America, Japan, EEC, and other industrialised nations. For each, differences in trade patterns have to be considered. The details are not needed for the present objective of suggesting that Calarco's forecast is consistent with steel industry consensus.

5. In viewing these comparisons, some problems of my calculations should be noted. IEA does not report coking coal use as such. My figures are derived as coking coal output less coking coal exports plus coking coal imports. Nonelectric steam coal then is computed as

total coal use less electric power use and calculated coking coal use. One clear drawback is the IEA trade figures aggregate coke exports and imports into an "other solid fuel" category. My calculations include these in steam coal trade. The deduction of coke exports and exclusion of any imports or output reported as coking coal but used for steam coal produce a downward bias in my calculation of steam coal consumption. To examine the impacts, I used my methodology to compute 1983 consumption figures from the data in the *Coal Information 1985* report. These then were compared to the balance sheet (and coking coal tonnage) data from OECD historical data reports as tabulated in Table 3.2. The biggest difference was in OECD Europe. The figures computed from *Coal Information* exceeded coking coal consumption tonnage by about 8.5 million TCE. German figures were too high by more than 12 million tonnes. British and French too low by more than 2 million TCE each. The Japanese coking figure was 3.5 million TCE too high. Equal and opposite errors arose in estimates of nonutility steam coal. As a rough adjustment for this problem, the predicted change in coking coal and nonutility steam coal demand from 1983 to 2000 was computed. Then forecasts were calculated as the sum of 1983 consumption as reported in the historical data reports and the change. The rationale is that the error in my method will remain roughly constant over time.

The IEA made several important changes in the substance and form of the coal forecasts issued in 1986. The key substantive change is a halving of forecasts of US coal exports (and moderately raising those for Australia) so that my complaint about overoptimism on exports no longer applies. The key formal change was the removal of the production and consumption of other solid fuels such as peat and wood wastes from the analysis. The prior reports had separated the historic and forecast levels of the production of these other solids but only gave historic figures on consumption. These last suggest that production equals consumption and predominantly occurs in what I have termed other steam uses. The US accounts for a larger historic share of these other solids than for solids as a whole and, more critically, was predicted to account for over 90 percent of the increase in use of these other solids – 54 million out of 60 million TCE. The predicted changes in production of other solids elsewhere in the OECD are scattered (and in some cases negative) and involve far fewer tonnes than in the US case. However, other solid fuel constituted almost all solid fuel output of six countries (Denmark, Finland, Ireland, Portugal, Sweden, and Switzerland) and so, that output and its change disappears from the forecasts for these countries.

To compare IEA views of the rise of other steam coal uses of coal, an adjustment must be made to remove other solids from the earlier other steam coal uses forecast. It turns out that clearcut downward adjustments were made for three leading Western European countries – France, Italy, and Germany. The 1986 forecasts set the respective changes at 8, 2, and 0 million TCE. In every case, examination of the details suggests that these are clearly reductions in the forecasts of use of hard coal and lignite in the three countries, since other solids use was not expected to grow significantly.

Bibliography

Ackerman, Bruce A., and William T. Hassler, 1981, *Clean Coal/Dirty Air.* New Haven: Yale University Press

Adelman, M. A., 1970, "Economics of exploration for petroleum and other minerals." *Geoexploration* 8, pp. 131–50

1972, *The World Petroleum Market.* Baltimore, Md.: Johns Hopkins University Press for Resources for the Future

Adelman, M. A., John C. Houghton, Gordon Kaufman, and Martin B. Zimmerman, 1983, *Energy Resources in an Uncertain Future, Coal, Gas, Oil, and Uranium Supply Forecasting.* Cambridge, Mass.: Ballinger Publishing Company

Alsworth, J., 1984, "Coal." *Canadian Mining Journal* (February), pp. 87–89.

American Iron and Steel Institute, annual, *Annual Statistical Report.* Washington, DC

Association Technique de l'Importation Charbonnière, annual, *Rapport Annuel*

Australian Joint Coal Board, annual, *Black Coal in Australia.* Sydney: Joint Coal Board

Barnett, Donald W., 1982, "The importance of coal to the future of the Australian economy." *Australian Coal Miner* (December), pp. 4–8, 66–68

1984/85, "Rail freight and the cost of Australian and North American coal." *Journal of Business Administration*, 15, pp. 175–201

1985a, "Export coal costs in Australia, Canada, South Africa and the USA." *Materials and Society*, 9:4, pp. 461–78

1985b, "The Australian export coal industry in the mid-1980s – an overview." *Materials and Society*, 9:4, pp. 443–60

1985c, *The Supply of USA Thermal Coal Export.* London: IEA Coal Research

Barnett, Harold J., and Chandler Morse, 1963, *Scarcity and Growth, the Economics of Natural Resource Availability.* Baltimore, Md.: Johns Hopkins University Press for Resources for the Future

Bauer, Christian, 1984, "Finanzpolitik und Energie." *Zeitschrift für Energiewirtschaft*, no. 4, pp. 255–61

Bending, Richard, 1981, *UK industrial Fuel Use, 1972–1979.* Cambridge: Cambridge Energy Research Group, discussion paper 17

1982, *A Revised Simulation Model of UK Industrial Energy Use.* Cambridge: Cambridge Energy Research Group, discussion paper 20

Bierman, Harold, Jr., and Seymour Smidt, 1975, *The Capital Budgeting Decision*, Fourth Edition. New York: Macmillan Publishing Company

British Petroleum Company, annual since 1982, *BP Statistical Review of World Energy*. London

annual to 1981, *BP Statistical Review of the World Oil Industry*. London

Calarco, Vincent J., Jr., 1985, *World Coal Outlook: A Reassessment*. New York: Chase Manhattan Bank

Canadian Mining Journal, 1983, Special Issue on Denison Mines (November)

Cattell, R. K., 1983, *Non Premium Fuel Markets in UK Industry: Prospects for Coal*. Cambridge: Cambridge Energy Research Group, discussion paper 25

Charbonnages de France, annual, *Rapport d'Activité* (Prior to report for 1981: *Rapport de Gestion*). Paris

annual, *Statistique Annuelle*. Paris

Coase, Ronald H., 1937, "The nature of the firm." *Economica*, N 8:4, pp. 386–405

1960, "The problem of social costs." *Journal of Law and Economics*, pp. 1–44

Comptoir Belge des Charbons, annual to 1982, *Statistics de base de l'industrie charbonnière*. Brussels

Crick, Michael, 1985, *Scargill and the Miners*, second edition. Harmondsworth: Penguin Books

Deam, R. J., and Carlos Giesoket, 1985, "Towards three laws of energy substitution" in Paul Tempest, ed. *Energy Economics in Britain*. London: Graham and Trotman, pp. 253–60

Düngen, Helmut, 1984, "Subventionen in der deutschen Energiewirtschaft von 1979 bis 1984." *Zeitschrift für Energiewirtschaft*, no. 4, pp. 262–9

Düngen, Helmut, and Dieter Schmitt, 1983, "Neuere Entwicklung und Perspektiven der Kohlesubventionierung in der Bundesrepublik Deutschland." *Zeitschrift für Energiewirtschaft*, no. 4, pp. 276–85

Dutton, C. M. J., 1982, *Modelling the International Trade in Steam Coal*. Cambridge: Cambridge Energy Research Group, discussion paper 21

Eden, Richard, Nigel Evans, and Roy Cattell, 1985, *World Coal: An Aide-Memoir*. Cambridge: Cambridge Energy Research Group, discussion paper 33

Edison Electric Institute, annual, *Annual Electric Power Survey*. Washington, DC: Edison Electric Institute

annual, *Statistical Yearbook of the Electric Utility Industry*. Washington, DC: Edison Electric Institute

1974, *Historical Statistics of the Electric Utility Industry through 1970*. New York: Edison Electric Institute

Electricité de France, annual, *EDF* (English language version of annual report). Paris

Ente Nazionale Per L'Energia Elettrica, annual, *Produzione e Consumo Di Energia Elettrica in Italia*. Rome

annual, *Report of the Board of Directors*. Rome

1983, *ENEL Programmes*. Rome

European Communities Commission, annual, *Memorandum on the financial aids granted by the Member States to the coal industry*. Brussels

annual, "The market for solid fuels in the Community and the outlook." *Official Journal of the European Communities*

1977, *Twenty-five Years of the Common Market in Coal*. Brussels

1982, *Review of Member States' Energy Policy Programmes and Progress towards 1990 Objectives*. Brussels

1983a, *Proposal for a Council Directive on the Limitation of emissions of pollutants from the air from large combustion plants*. Brussels

1983b, *Report on the Brown Coal and Peat Industries in the European Communities*. Brussels

1983c, *The New System for Coking Coal and Coke for the Iron and Steel Industry in the Community*. Brussels

1984, *Review of Member States' Energy Policies*. Brussels

1985a, *Amended Proposal for a Council Directive on the limitation of emissions of pollutants from the air from large combustion plants*. Brussels

1985b, *Energy 2000, a reference projection and its variants for the European Community and the World to the Year 2000*. Luxembourg

European Communities, Statistical Office, annual, *Basic Statistics of the Community, Comparisons with some European countries, Canada, the USA, Japan and the USSR*. Luxembourg

annual, *Energy Statistics Yearbook*. Luxembourg

annual, *Iron and Steel Yearbook*. Luxembourg

monthly, *Coal*. Luxembourg

monthly, *Electric Energy*. Luxembourg

monthly, *Hydrocarbons*. Luxembourg

monthly, *Iron and Steel*. Luxembourg

Evans, Nigel, and Chris Hope, 1984, *Nuclear Power: Futures, Costs and Benefits*. Cambridge: Cambridge University Press

Evans, Nigel, and Hadi Dowlatabadi, 1985, *Trading Power across Frontiers*. Cambridge: Cambidge Energy Research Group, discussion paper 32

Exxon Coal International and Intercor, Colombia, 1984, "The Cerrejón project." *Bulk Solids Handling*, 4:2 (June)

Fédération Charbonnière de Belgique, annual since 1983, *Statistiques charbonnières simplifiées*. Brussels

monthly, *Bulletin Statistique Mensuel*. Brussels

Fisher, Irving, 1930, *Theory of Interest*. New York: Macmillan Publishing Company

Gaskin, Maxwell, 1981, *Market Aspects of an Expansion in the International Steam Coal Trade*. London: IEA Coal Research

1983, *Organisation and Structure of the Pacific Steam Coal Trade*. London: IEA Coal Research

Germany, Federal Republic, Bundesministerium für Wirtschaft, annual, *Daten zur Entwicklung der Energiewirtschaft in der Bundesrepublik Deutschland*. Bonn

Glyn, Andrew, 1984, *The Economic Case Against Pit Closures*. Sheffield: National Union of Mineworkers

Gordon, Richard L., 1966, "Conservation and the theory of exhaustible resources." *The Canadian Journal of Economics and Political Science*, 32:3(August), pp. 319–26

1970, *The Evolution of Energy Policy in Western Europe: The Reluctant Retreat from Coal*. New York: Praeger Publishers

1975, *US Coal and the Electric Power Industry*. Baltimore, Md.: Johns Hopkins University Press for Resources for the Future

1978, *Coal in the US Energy Market: History and Prospects*. Lexington, Mass.: D. C. Heath Lexington Books

1981, *An Economic Analysis of World Energy Problems*. Cambridge, Mass.: The MIT Press

1982, *Reforming the Regulation of Electric Utilities: Priorities for the 1980s*. Lexington Mass.: D. C. Heath Lexington Books

Grainger, L., and L. Gibson, 1981, *Coal Utilisation Technology, Economics and Policy*. London: Graham & Trotman

Hewett, Ed A., 1984, *Energy Economics and Foreign Policy in the Soviet Union*. Washington DC: The Brookings Institution

Hirschleifer, J., 1970, *Investment, Interest, and Capital*. Engelwood Cliffs, N.J.: Prentice-Hall

Holcomb, Robert S., and Mike Prior, 1985, *Economics of Coal for Steam Raising in Industry*. London: IEA Coal Research

Holmes, J. M., D. F. Hemming, and M. Teper, 1984, *The Cost of Liquid Fuels from Coal: Part II Fischer-Tropsch Liquids*. London: IEA Coal Research

Holschuh, Lenhard J., 1984, *Annual Report of the Secretary General (of the International Iron and Steel Institute)*. Brussels: International Iron and Steel Institute

Houillères de Bassin du Centre et du Midi, annual, *Rapport d'Activité* (formerly *Rapport de Gestion*). St. Etienne

Houillères du Bassin de Lorraine, annual, *Rapport d'Activité* (formerly *Rapport de Gestion*). Freyming-Merlebach

Houillères du Bassin du Nord et du Pas-de-Calais, annual, *Rapport d'Activité* (formerly *Rapport de Gestion*). Douai

Hughes, P. R., 1982, *CEGB Proof of Evidence on Fossil Fuel Supplies*. London: UK Central Electricity Generating Board

International Bank for Reconstruction and Development (the World Bank), 1983, *China: Socialist Economic Development*, v. 2 The Economic Sectors. Washington, DC

1985, *China: Long-Term Issues and Options*, Annex C Energy. Washington, DC

International Energy Agency, annual, *Coal Information*. Paris: Organisation for Economic Co-operation and Development

annual, *Energy Balances of OECD Countries*. Paris: Organisation for Economic Co-operation and Development

annual, *Energy Policies and Programmes of IEA Countries*. Paris: Organisation for Economic Co-operation and Development

annual, *Energy Statistics and Main Historical Series*. Paris: Organisation for Economic Co-operation and Development

biennial, *Coal Prospects and Policies in IEA Countries*. Paris: Organisation for Economic Co-operation and Development

1978, *Steam Coal: Prospects to 2000*. Paris: Organisation for Economic Co-operation and Development

1982a, *The Use of Coal in Industry*. Paris: Organisation for Economic Co-operation and Development

1982b, *World Energy Outlook*. Paris: Organisation for Economic Co-operation and Development

1984, *Energy Balances of Developing Countries 1971–1982*. Paris: Organisation for Economic Co-operation and Development

1985, *Electricity in IEA Countries Issues and Outlook*. Paris: Organisation for Economic Co-operation and Development

International Iron and Steel Institute, annual, *Steel Statistical Yearbook*. Brussels

Keeny, Spurgeon, M., Jr. (chairman, Nuclear Energy Policy Study Group), 1977, *Nuclear Power Issues and Choices*. Cambridge, Mass.: Ballinger Publishing Company

Keystone Coal Industry Manual, annual, *Keystone Industry Manual*. New York: McGraw-Hill

Landsberg, Hans H. (study group chairman), 1979, *Energy: The Next Twenty Years*. Cambridge, Mass.: Ballinger Publishing Company

Lave, Lester, and Eugene P. Seskin, 1977, *Air Pollution and Human Health*. Baltimore, Md.: Johns Hopkins University Press for Resources for the Future

Lee, Hugh M., 1982a, *The Future Cost and Availability of Thermal Coal Exports from Australia*. London: IEA Coal Research

1982b, *The Future Cost and Availability of Thermal Coal Exports from Canada*. London: IEA Coal Research

1982c, *The Future Cost and Availability of Thermal Coal Exports from South Africa*. London: IEA Coal Research

Lister, Louis, 1960, *Europe's Coal and Steel Community*. New York: Twentieth Century Fund

Long, Ray, 1982, *Constraints on International Trade in Coal*. London: IEA Coal Research

Lund, Robert C., Kenneth J. Arrow, Gordon R. Corey, Partha Dasgupta, Amanta Sen, Thomas Stauffer, Joseph E. Stiglitz, J. A. Stockfisch, and Robert Wilson, 1982, *Discounting for Time and Risk in Energy Policy*. Baltimore, Md.: Johns Hopkins University Press for Resources for the Future

Marshall, Eileen, and Colin Robinson, 1984, *The Economics of Energy Self-Sufficiency*. London: Heinemann Educational Books

Merrick, D., 1984, *Coal Combustion and Conversion Technology*. London: Macmillan Publishers, Ltd., and New York: Elsevier Science Publishing Company

Merritt, A. K., and Allen Sykes, 1973, *The Finance and Analysis of Capital Projects*, second edition. London: Longman Group, and New York: John Wiley & Sons

Merritt, Paul C, 1983, "Cerrejón puts Columbia in the spotlight." *Coal Age* (November), unpaged reprint

National Coal Association, annual, *Coal Data*. Washington, DC

annual, *Coal Traffic*. Washington, DC

annual, *International Coal*. Washington, DC

annual, *Steam Electric Plant Factors*. Washington, DC

1960, *Trends in Electric Utility Industry Experience, 1946–1958*. Washington, DC

1972, *World Coal Trade 1972 Edition*. Washington, DC

1985, *Power Plant Coal Deliveries Cost & Quality*. Washington, DC

Navarro, Peter, 1985, *The Dimming of America, the Real Costs of Electric Utility Regulatory Failure*. Cambridge, Mass.: Ballinger Publishing Company

Nelson, Robert H., 1983, *The Making of Federal Coal Policy.* Durham, NC: Duke University Press

North American Electric Reliability Council, annual, *Electric Power Supply and Demand.* Princeton, NJ

NUS Inc., 1977a, *Coal Mining Costing Models – Underground Mines.* Palo Alto: Electric Power Research Institute

1977b, *Coal Mining Costing Models – Surface Mines.* Palo Alto: Electric Power Research Institute

Phillips, Keri, 1984, "El Cerrejón coal – early coal project reschedules export start for April 1985." *International Bulk Journal* (March), unpaged reprint

Queensland Coal Board, annual, *Annual Report.* Brisbane

Ramsay, William, 1979, *Unpaid Costs of Electric Energy, Health and Environmental Impacts from Coal and Nuclear Power.* Baltimore, Md.: Johns Hopkins University Press for Resources for the Future

Roberts, Peter W., 1985, "Energy policy in the United Kingdom: the case of coal and regional development." Unpublished manuscript

Robinson, Colin, and Eileen Marshall, 1981, *What Future for British Coal? Optimism or Realism on the Prospects up to the Year 2000.* London: The Institute of Economic Affairs

Ruhrkohle AG, annual, *Annual Report.* Essen: Ruhrkohle Aktiengesellschaft

Saarberg, Zentrale Unternehmensplanung, annual, *Marktperspektiven.* Saarbruchen

Samuelson, Paul A., 1954, "The pure theory of public expenditure," *Review of Economics and Statistics* 36:4, reprinted in Joseph E. Stiglitz, ed., *The Collected Scientific Papers of Paul A. Samuelson,* V. 2. Cambridge, Mass.: The MIT Press, pp. 1223–25

1955, "Diagramatic exposition of a theory of public expenditure," *Review of Economics and Statistics* 37:4, reprinted in Joseph E. Stiglitz, ed., *The Collected Scientific Papers of Paul A. Samuelson,* V. 2. Cambridge, Mass.: The MIT Press, pp. 1226–32

1958, "Aspects of public expenditure theories," *Review of Economics and Statistics* 40:4, reprinted in Joseph E. Stiglitz, ed., *The Collected Scientific Papers of Paul A. Samuelson,* V. 2. Cambridge, Mass.: The MIT Press, pp. 1233–39

1967, "Indeterminacy of governmental role in public-good theory," *Papers on Non-Market Decision Making,* reprinted in Robert C. Merton, ed., *The Collected Scientific Papers of Paul A. Samuelson,* V. 3, Cambridge, Mass.: The MIT Press, p. 521

1968, "Pitfalls in the analysis of public goods," *Journal of Law and Economics* (January), reprinted in Robert C. Merton, ed., *The Collected Scientific Papers of Paul A. Samuelson,* V. 3, Cambridge, Mass.: The MIT Press, pp. 522–27

1969, "Pure theory of public expenditure and taxation," in J. Margolis and H. Guitton, eds., *Public Economics,* London: Macmillan, reprinted in Robert C. Merton, eds., *The Collected Scientific Papers of Paul A. Samuelson,* V. 3, Cambridge, Mass.: The MIT Press, pp. 492–517

Schmidt, Richard A., 1979, *Coal in America: An Encyclopedia of Reserves, Production and Use.* New York: McGraw-Hill Coal Week

Skea, J. F., 1981a, *Converting from Oil to Coal Firing: A Case Study at a Large Industrial*

Site. Cambridge: Cambridge Energy Research Group, discussion paper 4

1981b, *Modelling Coal Penetration in the Industrial Steam Raising Markets: An Engineering Approach*. Cambridge: Cambridge Energy Research Group, discussion paper 6

1981c, *The Prospects for a Revival of Coal Use in UK Industry*. Cambridge: Cambridge Energy Research Group, discussion paper 13

1983, "Reviving industrial coal use: a case study of the United Kingdom. *Energy Systems and Policy* (7:3), pp. 91–113

1985, *A Simulation Model of Interfuel Substitution in the Industrial Boiler Market*

Smith, V. Kerry, ed., 1979, *Scarcity and Growth Reconsidered*. Baltimore, Md.: Johns Hopkins University Press for Resources for the Future

Sorensen, Jean, 1985, "Contracts for BC coal." *Canadian Mining Journal* (June), pp. 20–27

South Africa Chamber of Mines, annual, *Statistical Tables*. Johannesburg: Chamber of Mines

South Africa, Republic of, annual, *Report of the Department of Mineral and Energy Affairs*. Pretoria

Speight, James G., 1983, *The Chemistry and Technology of Coal*. New York: Marcel Dekker Inc.

Statistik der Kohlenwirtschaft, annual, *Der Kohlenbergbau in der Energiewirtschaft der Bundesrepublik Deutschland*. Essen: Statistik der Kohlenwirtschaft

semiannual, *Zahlen zur Kohlenwirtschaft*. Essen: Statistik der Kohlenwirtschaft

Steenblik, R. P., 1985, *Issues in Modelling International Coal Supply*. Rotterdam: Centre for International Energy Studies, Erasmus University

Stern, Jonathan P., 1982, *East European Energy and East–West Trade in Energy*. London: Policy Studies Institute

1984, *International Gas Trade in Europe: the Politics of Exporting and Importing Countries*. London: Heinemann Educational Books

1985, *Gas's Contribution to UK Self-sufficiency*. London: Heinemann Educational Books

Stigler, George J., 1968, "The division of labor is limited by the extent of the market," in George J. Stigler, *The Organization of Industry*. Homewood: Richard D. Irwin, pp. 129–41

Sugden, Robert, 1984, "Reciprocity: the supply of public goods through voluntary contributions." *The Economic Journal*, (94/December), pp. 772–87

UK Central Electricity Generating Board, annual, *Annual Report & Accounts*. London: UK Central Electricity Generating Board

1982a, *UK Coal Costs and Prices*. London: UK Central Electricity Generating Board

1982b, *World Coal Supply*. London: UK Central Electricity Generating Board

1982c, *World Energy Background to 2030*. London: UK Central Electricity Generating Board

1985, *Closing Submission – Need and Economics, Chapters 9–20*. London: UK Central Electricity Generating Board

UK Department of Energy, annual, *Digest of United Kingdom Energy Statistics*

(Responsible agency and title altered several times). London: Her Majesty's Stationery Office

UK Department of Energy (R. J. Priddle), 1982, *Proof of Evidence for the Sizewell 'B' Public Inquiry*. London: UK Department of Energy

UK House of Lords, Select Committee on the European Communities, 1984, *Air Pollution*. London: Her Majesty's Stationery Office

UK Monopolies and Mergers Commission, 1983, *National Coal Board: A Report on the Efficiency and Costs in the Development, Production and Supply of Coal by the NCB*. London: Her Majesty's Stationery Office

UK National Coal Board, annual, *Reports and Accounts*. London

UK National Coal Board, South Midlands Area, 1985, *The South Warwickshire Project: A Consultation Paper*. London

UK South of Scotland Electricity Board, annual, *Reports and Accounts*. Glasgow

United Nations, annual, *Energy Statistics Yearbook*. New York

UN Economic Commission for Europe, annual, *The Coal Situation in the ECE Region and its Prospects*. Geneva

1983, *World Coal Trade up to the Year 2000*. Geneva

US Bureau of Mines, annual, *Minerals Yearbook*. Washington, DC: US Government Printing Office

US Central Intelligence Agency, 1985, *USSR Energy Atlas*

US Commission on Fair Market Value Policy for Federal Coal Leasing, 1984, *Report of the Commission: Fair Market Value Policy for Federal Coal Leasing*. Washington, DC: US Government Printing Office

US Congress, Joint Economic Committee, 1981, *East European Economic Assessment*. Washington, DC: US Government Printing Office

1982, *China Under the Four Modernizations*. Washington, DC: US Government Printing Office

1982, *Soviet Economy in the 1980s: Problems and Prospects*. Washington, DC: US Government Printing Office

US Congress, Office of Technology Assessment, 1978. *A Technology Assessment of Coal Slurry Pipelines*. Washington, DC: US Government Printing Office

1979, *The Direct Use of Coal: Prospects and Problems of Production and Combustion*. Washington, DC: US Government Printing Office

1981a, *An Assessment of Development and Production Potential of Federal Coal Leases*. Washington, DC: US Government Printing Office

1981b, *Technology & Soviet Energy Availability*. Washington, DC: US Government Printing Office

1984a, *Acid Rain and Transported Air Pollutants: Implications for Public Policy*. Washington, DC: US Government Printing Office

1984b, *Environmental Protection in the Federal Coal Leasing Program*. Washington, DC: US Government Printing Office

US Department of Energy, Energy Information Administration, annual, *Annual Energy Outlook*. Washington, DC: US Government Printing Office

annual since 1982, *Annual Energy Review*. Washington, DC: US Government Printing Office annual to 1981, *Annual Report to Congress*, Vol. 2 Energy Statistics, Vol. 3 Energy Projections. Washington, DC: US Government Printing Office

annual, *Coal Production*. Washington, DC: US Government Printing Office

annual, *Cost and Quality of Fuels for Electric Utility Plants*. Washington, DC: US Government Printing Office

annual, *Electric Power Annual*. Washington, DC: US Government Printing Office

annual since 1982, *Financial Statistics of Selected Electric Utilities*. Washington, DC: US Government Printing Office

annual since 1981, *Historical Plant Construction Cost and Annual Production Expenses for Selected Electric Plants*. Washington, DC: US Government Printing Office

annual, *International Energy Annual*. Washington, DC: US Government Printing Office

annual, *Inventory of Power Plants in the United States*. Washington, DC: US Government Printing Office

annual, *State Energy Data Report*. Washington, DC: US Government Printing Office.

annual to 1981, *Statistics of Privately Owned Electric Utilities (Classes A and B Companies)*. Washington, DC: US Government Printing Office

annual to 1978, *Steam-Electric Plant Construction Cost and Annual Production Expenses*. Washington, DC: US Government Printing Office

annual (1979 and 1980), *Thermal-Electric Plant Construction Cost and Annual Production Expenses*. Washington, DC: US Government Printing Office

monthly, *Monthly Energy Review*. Washington, DC: US Government Printing Office

quarterly, *Coal Distribution*. Washington, DC: US Government Printing Office

quarterly, *Quarterly Coal Report*. Washington, DC: US Government Printing Office

weekly, *Weekly Coal Production*. Washington, DC: US Government Printing Office

US Department of Interior, Bureau of Land Management, 1979, *Final Environmental Statement Federal Coal Management Program*. Washington, DC: US Government Printing Office

US Public Land Law Review Commssion, 1970, *One Third of the Nation's Land*. Washington, DC: US Government Printing Office

Unternehmensverband Ruhrbergbau, 1955, *Die Kohlenwirtschaft der Welt in Zahlen*. Essen: Verlag Glückauf

1961, *Die Kohlenwirtschaft der Welt in Zahlen*. Essen: Verlag Glückauf

Verein Deutscher Kohlenimporteure, annual, *Jahresbericht*. Hamburg: Verein Deutscher Kohlenimporteure

Vereinigung Industrielle Kraftwirtschaft, annual, *Statistik der Energiewirtschaft*. Essen: Vereinungung Industrielle Kraftwirtschaft

Verlag Glückauf, annual, *Jahrbuch für Bergbau, Energie, Mineralöl und Chemie*. Essen: Verlag Glückauf

Williamson, Oliver E., 1975, *Market and Hierarchies: Analysis and Antitrust Implications*. New York: The Free Press

Wilson, Carroll L., Project Director, 1980, *Coal – Bridge to the Future, Report of the World Coal Study*. Cambridge, Mass.: Ballinger Publishing Company

Wilson, Richard, Steven D. Colome, John D. Spengler, and David Gordon Wilson, 1980, *Health Effects of Fossil Fuel Burning, Assessment and Mitigation*. Cambridge, Mass.: Ballinger Publishing Company

Wright, Andrew, 1985, "El Cerrejón moves into the market." *Coal Age* (October), pp. 69–72

Zimmerman, Martin B., 1981, *The US Coal Industry: The Economics of Policy Choice*. Cambridge, Mass.: The MIT Press

Index